Biological Treatment of Microbial Corrosion

Opportunities and Challenges

Biological Treatment of Microbial Corrosion

Opportunities and Challenges

REZA JAVAHERDASHTI, PhD
Director
ParsCorrosion
Perth, Australia

KIANA ALASVAND, PhD
ParsCorrosion
Perth, Australia

ELSEVIER

BIOLOGICAL TREATMENT OF MICROBIAL CORROSION ISBN: 978-0-12-816108-1

Publisher: Matthew Deans
Acquisition Editor: Christina Gifford
Editorial Project Manager: Peter Adamson
Production Project Manager: Kiruthika Govindaraju
Cover Designer: Alan Studholme

ELSEVIER

3251 Riverport Lane
St. Louis, Missouri 63043

Preface

Corrosion, in general, and microbial corrosion (MIC), in particular, are very significant issues that do harm to human-made assets and structures. Industry-wise, it is almost all industries that suffer from MIC and may include oil and gas, pipeline, water and waste water treatment, chemical plants, ports and jetties, ships, to name a few.

It may sound surprising to the reader that although MIC is so recognized, there are professionals who take it lightly and simply think that by the ways they have been using so far, it is possible to cure and control MIC.

Apart from a few exemplary universities and research institutes that have largely contributed to filling the gap between industry and academia with respect to MIC research and engineering, a majority of technological thinktanks—and therefore researchers—are not yet aware of MIC and how such a complex phenomena must be addressed and even predicted properly: if we accept that softwares are based on mathematical models, lack of specifically prepared softwares for MIC is itself a good indication of how poorly mathematics has been used to look at this forgotten corrosion process. To tackle this weakness, we have suggested a mathematical model based on fuzzy logic that we previously used with success for corrosion-related bacteria (CRB) to predict their corrosive actions. This modeling now aims to highlight how pathogenic bacteria can also be treated likewise.

In addition to being intrinsically complex because of accommodating various interactions between living things and nonliving things, microbial corrosion also suffers from inadequate and sometimes wrong terminology that has been in use for decades in its literature: an example of wrong terms used in MIC studies is the term "biofilm," we have suggested the term "Temenos" for that and we have explained our reasons for this suggestion although we have still used "biofilm" in majority of the material given in this book.

In this book, following focusing on "corrosion control" and "corrosion prevention," for the first time, we coined the terms "Zugzwang Effect" and "Myth Effect" and explained their importance especially from a microbial corrosion management point of view. These two terms are meant to add more meaning to what industry does with regard to recognition, treatment, and monitoring of microbial corrosion.

In addition to the above innovations in this book, we believe that the main innovation is still in its title: so far in all industries, MIC has been tried to be controlled by the aid of mechanical-physical (for example, coatings or pigging), electrical (cathodic protection), and chemical (use of biocides) means. This book will talk about a possibility that there can exist yet another means that is biological. In other words, how bacteria against bacteria can be used to control MIC.

We do hope that with these innovative points in this book, our readers from engineers to researchers will find it very useful, practical, and question-making. If after reading this book, the reader has come up with some questions to be answered in his/her head, we believe that we have reached our goals.

The structure of the book is that it has two sections in six chapters. The first section that contains three chapters has been authored by Dr. Reza Javaherdashti. The second section that contains three chapters has been written and prepared by Dr. Kiana Alasvand.

We would like to dedicate this book to our beloved ones, Helya and Hannah Javaherdashti, Tahmoures Alasvand, and Mahin Soleimani.

Let us close this chapter with a verse from a great man, the Iranian mathematician, astronomer, and poet, Omar Khayyam who said:

هرگز دل من ز علم محروم نشد / کم ماند ز اسرار که معلوم نشد

هفتاد و دو سال فکر کردم شب و روز / معلومم شد که هیچ معلوم نشد

Never did I give up from learning/ To me remained unrevealed rarely secrets of the Universe Spent I seventy-two years of my life in thinking/ what I got as a result was that I knew nothing

Reza Javaherdashti, PhD
Kiana Alasvand, PhD

Contents

1 Why Corrosion and Particularly Microbial Corrosion Are Important?, *1*

The Definition of Engineering Importance as a Function of "Risk" and "Cost", *1*

What are the Risks and Costs of Corrosion and Microbial Corrosion, *2*

Economical Costs, 2

Ecologic Costs, 3

How can Risks, Defined as a Function of Likelihood and Consequences, be Related to Microbial Corrosion?, 6

References, *9*

2 A Brief Introduction to Corrosion Engineering, *11*

Introduction, *11*

The Thermodynamics of Corrosion and Its Importance in Dealing With It in Industries, *11*

Some Basic Concepts of Corrosion, *15*

Essentialities for Corrosion, *15*

"Series" and "Parallel" Corrosion Process Patterns, *16*

A Combination of a Concrete Slab and a Steel Rod, *16*

How is Corrosion Treated?, *17*

Application of Cathodic Protection Methods, *18*

Materials Selection, *19*

Application of Coatings, *20*

Design Factor, *21*

Chemical Treatment, *22*

Closing Remarks, *23*

References, *24*

3 An Introduction to Microbial Corrosion, *25*

How Are Microbial Corrosion, in Terms of Being an Electrochemical Process, and Microbiology Related?, *25*

What Is MIC and What Are the Four Basic Characteristics of It?, *25*

MIC as an Electrochemical Process, 25

Diversity of Organisms Involved in Microbial Corrosion, 26

Suitable Environment for MIC to Occur, 27

Biofilm, A True Concept With a Wrong Name, 27

Introducing Some Bacteria That Are Related to Corrosion (Corrosion-Related Bacteria = CRB), *30*

Sulfate-Reducing Bacteria, 32

Iron-Oxidizing Bacteria, 33

Iron-Reducing Bacteria, 34

Acid-Producing Bacteria—Nonorganic Producers, 34

Acid-Producing Bacteria—Organic Acid Producers, 34

Possible Role of Magnetic Bacteria in Inducing Corrosion, 35

Use of Mathematics in Microbiology: Fuzzy Modeling for Distinguishing CRB and Pathogenic Bacteria, *36*

Eight Fundamental Assumptions, 37

General Methodology, 37

CRB Fuzzy Modeling, *38*

Defining Membership Function for CRB, 38

PB Fuzzy Modeling, *39*

Physicoelectrostatic Factors, 40

Mechanical Factors, 40

Fuzzy Composite Functions for PB, 41

Discussion, *41*

Summary, *42*

How Is MIC Recognized? A Short Review of Culture-Dependent and Culture-Independent Methods and Their Pros and Cons, *42*

Culture-Dependent Methods, 42

Culture-Independent Methods, 45

"Zugzwang Effect" in MIC, *46*

"Myths Effect" in Recognition and Treatment of MIC, *52*

Javaherdashti Model for CM-MIC, *55*

Closing Remarks, *67*
References, *67*

4 An Introduction to Microbiology for Nonmicrobiologists, *71*
Introduction, *71*
Cell and Macromolecules, *71*
Amino Acids, 71
Fatty Acids, 73
Monosaccharides (or Simple Sugar), 74
Nucleotide, 74
Classification of Organisms, *78*
Bacterial Cell Architecture, *81*
Appendages, 81
Cell Envelope, 83
Cytoplasmic Region, 85
Microbial Genetic Material, *86*
Microbial Metabolism, *86*
Reproduction of Microorganism, *87*
Microbial Growth, *88*
Dormancy in Bacteria, *89*
Physical and Chemical Requirement for Bacterial Growth, *89*
Bacterial Signaling Pathways, *90*
Quorum-Sensing Signaling, 90
Cyclic Dinucleotide Signaling, 91
Bacterial Locomotion, *91*
Biofilm, *92*
Inhibition of Microbes, *93*
Microbial Interaction, *94*
References, *96*

5 Biologic Treatment of Corrosion, *101*
Introduction, *101*
Corrosion Inhibition by Predatory Microorganisms, *101*
Corrosion Inhibition by Predatory Bacteria, 101
Corrosion Inhibition by Predatory Phages, 107
Corrosion Inhibition by Nitrate-Reducing Bacteria, *113*
Corrosion Inhibition by Regenerative Biofilm, *116*
Corrosion Inhibition Through the Removal of Corrosive Agents, *118*
Corrosion Inhibition Through a Protective Layer, *119*
Corrosion Inhibition Through Antimicrobial Producing Biofilm, *119*
Corrosion Inhibition With Biologically Produced Corrosion Inhibitor, *120*
Corrosion Inhibition With Biosurfactant, *120*
Corrosion Inhibition by Biosynthesized Nanoparticles, *122*
Corrosion Inhibition by Hydrophobic Biofilm, *123*
Corrosion Inhibition by Microbial Antifreeze Protein, *125*
Inhibition of Corrosion by Targeting Signaling Molecule, *127*
Corrosion Inhibition by Quorum Sensing Inhibitors, 127
Inhibition of Corrosion by Targeting c-di-GMP, 130
Corrosion Inhibition by Microbial Footprint, *130*
Corrosion Inhibition by Microbial Cell Surface Display, *132*
References, *133*

6 Future Perspective of Biological Inhibition of Corrosion, *145*
References, *148*

INDEX, *151*

CHAPTER 1

Why Corrosion and Particularly Microbial Corrosion Are Important?

THE DEFINITION OF ENGINEERING IMPORTANCE AS A FUNCTION OF "RISK" AND "COST"

Anything can be specified with two variables: its importance (the role that it plays in our lives) and the cost of being at this status. A rather trivial example could be our pets: the role that these lovely creatures may play in the lives of many of us is hard to deny. On the other hand, the cost for looking after them is a figure that many won't (can't) afford. Of course here we don't mean economy for keeping a pet, I am sure many of us have experienced the rather bitter feeling when we have read "No Dogs Allowed" signs or even the trauma that may be felt when a loved pet is not with us anymore. All these can make something, from a pet to corrosion, become important for us.

However, the importance of anything is rather a relative measure: something that may be important for someone may not be important for another person. Engineering importance, on the other hand, is another story. In fact, engineering importance because of its very nature cannot vary from person to person. It has a clear, mathematical definition that can be stated as below:

$$\text{Engineering Importance} = \text{Cost} \times \text{Risk}$$

Each of these terms has its own nonvarying, definitive meaning as we shall explain below in a moment. However, we have to add a very significant point here too: if engineering importance of something is important, no excuse is acceptable for ignoring it. Although this statement may sound obvious, our real life experience shows that in many cases in industries, the engineering importance of corrosion is taken not with the gravity it deserves.

When it comes to corrosion, we, as H.H. Uhlig did, can categorize corrosion losses as below:

1. Waste of energy and materials
2. Economical loss
   a. Direct loss
   b. Indirect loss
      i. Shutdown
      ii. Loss of efficiency
      iii. Product contamination
      iv. Overdesign

We will very briefly just touch economic-ecologic aspects of corrosion loss. However, this is not to be understood that other aspects of corrosion loss can be ignored. Take, for example, the overdesign that will be imposed to compensate for corrosion.

Assume that we have a 32-inch pipeline with a length of 10 km. If the thickness of the pipe is 5 mm, the internal volume of the pipe will be 5064.0 m^3. However, if we could manage to decrease this overdesign to 3 mm, the internal volume will be increased to 5114.0 m^3, in other words the internal volume will be increased by more than 90%.

The above shows that the overdesigning to be protected against corrosion is not an engineering solution. In addition to decreasing the useful volume of the pipe, overdesigning can indirectly add up more into the pollution: overdesign as a measure of protection against corrosion will add into the "embedded energy" (EmE). Embedded energy is defined as *"the energy consumed by all of the processes associated with the production of a structure from the acquisition of natural resources to product delivery."*[2] Therefore, when we do overdesign, we are augmenting the energy required to make the part (in our example, the 10-km, 32-inch pipe).

Biological Treatment of Microbial Corrosion. https://doi.org/10.1016/B978-0-12-816108-1.00001-X

This augmented energy will consume up more energy from extraction and refining of the metal to melting and shaping and deformation. Just multiply this figure by the total length of pipelines used around the world and those that are designed not to pose any corrosion risk by applying overdesign.

The example above can be taken as a standalone case of how corrosion can affect our lives and energy sources. Below, we will briefly introduce some economic-ecologic disadvantages of corrosion.

WHAT ARE THE RISKS AND COSTS OF CORROSION AND MICROBIAL CORROSION

We will start with what we mean by "cost" in the definition of engineering importance. In the context of this chapter, cost will mean anything that will increase the likelihood of creating a hazard to our investment. When we build a power plant or a refinery, for example, we do it by considering if this investment is feasible enough for us. In other words, we study the items required to ensure us that we are not "wasting our money." One important parameter here is the cost that is imposed on our investment—that is, the power plant or refinery or pipeline grid we make. The cost here can be categorized into two sets: the intrinsic costs associated with the pass of time and effect of elements on our investment (that can roughly be addressed as "depreciation" factor) and the costs that our investment may impose on the surrounding environment (ecologic costs). We will describe the depreciation under "economic costs," bearing in mind that depreciation in its very financial meaning does not include corrosion properly, as we will discuss it below.

Economical Costs

Whenever we do an investment, we are also obliged to think how long this investment will last. Especially if the investment is in a physical form (for instance, building up a power plant, a refinery, or the like), we are also interested to know how long we can expect this investment will last as well as after what period in time, this investment will start to become economically feasible. All physical assets we have in industry are exposed to the elements and thus will go through a process known by economists as "erosion": the value of your initial investment will depreciate by many factors such as "wear and tear." If we can assume a certain rate of annual depreciation/wear and tear for a physical asset, then it will be possible to also calculate from what point in time the asset will start to become economically feasible (this point in time is the "break-even point").

However, there are at least two shortcomings in such economical considerations:

1. Wear and tear—as the name also implies—can only contain physical forms of corrosion such as but not limited to erosion-corrosion or perhaps rusting (in iron and steel alloys). The term is not inclusive of chemical, electrochemical, and biologic factors that can affect corrosion and bring it to the final stage of having lost its mechanical integrity and strength.
2. Although the depreciation rate may be assumed certain, constant value for each year, as it is not considering the wider concept of corrosion, the actual depreciation rates to be imposed on the material will be much higher than what is expected on the paper. The main reason behind intensifying deprecation rates is corrosion.

Having said the above, it is now evident that as long as there is no mutual understanding between corrosion experts and economists, what is being reported as "cost of corrosion" must be taken as the very minimum values that could have been surveyed and calculated.

What is very interesting about the countries mentioned in Table 1.1 is that by proper management of cost of corrosion, they have managed to lower this cost. Perhaps the most striking example in this regard will be Kuwait: in 1995 the cost of corrosion has been 5.2% of its GDP. This cost of corrosion had actually been inherited from another study that had been done for this country in 1987, yielding the same figure of 5.2% GDP for the cost of corrosion.[3] However, by seriously

TABLE 1.1
Cost of Corrosion as a Percentage of Gross National Product (GNP) or Gross Domestic Product (GDP)[1] (used with permission)

Country	Year	Cost	Percentage of the GNP (%)
USA	1949	USD $5.5 billion	2.1
West Germany	1969	USD $6 billion	3
UK	1971	£1365 billion pounds	3.5
Australia	1972	AUD $900 million	3.5
Australia	1974	AUD $470 million	1.5
Japan	1977	USD $9.2 billion	1.8
USA	1978	USD $70 billion	4.5(GDP)
Australia	1983	AUD $2 billion	1.5
Kuwait	1995	USD $1 billion	5.2(GDP)
USA	1998	USD $276 billion	3.1(GDP)
Japan	1999	3.9 trillion Yen (Uhlig), 5.3 trillion Yen (Hoar)	0.77 (Uhlig), 1.02 (Hoar)
		9.7 trillion Yen (input/output)	1.88 (input/output)

working on lowering this cost, another study on the cost of corrosion in Kuwait in 2011 revealed that this cost had been decreased to the splendid figure of 1.7% GDP.[3] Now, if we consider that the cost of MIC is normally 20%[4] to 40%[5] of the total cost of corrosion, then it is easy to calculate the share of cost because of MIC in a country's economy. Javaherdashti[6] has reviewed some examples of MIC costs in various industries and countries.

Ecologic Costs

To understand the link between MIC and environment, one has to first understand the link between corrosion and environment because without appreciating the existence of such a link, there will no meaning in mentioning the ecologic importance of MIC. First, we have to define what we mean by "corrosion effect" and "corrosion impact".[2]

- **Corrosion effects** address those natural causes of corrosion that can be expressed by corrosion-related terminology and within a corrosion-related context.
- **Corrosion impacts** will address those natural results of corrosion that can be expressed in a quantitative way in terms of economical costs and measurable ecologic/community risks.

To understand the interaction between corrosion and environment, we will define two types of ecologic impacts of corrosion[2]:

1. Impacts concerning the corrosion of a particular system in a given industry (*corrosion of system impacts* [CSIs])
2. Impacts concerning use of corrosion countermeasures on a given system in a given industry (*counter-corrosion impacts* [CCIs])

As an example, although in a particular case of a pipeline explosion, the *effect* could be stress corrosion cracking (SCC), the *impacts* could range from the high cost of material and labor to serious ecologic dimensions such as leakage of the toxic fluid in the pipeline into the surrounding environments, thus causing pollution.

One important aspect of CSIs is the algorithm by which this impact can be evaluated. The evaluation procedure (algorithm) for CSIs is as follows:

1. A materials approach toward the system of interest. In other words, what are the materials from which the system has been manufactured.
2. Standards that have been used in manufacturing and corrosion management of the system such as materials selection, hydrotesting, coating selection criteria, etc.

3. Possible corrosion mechanisms and scenarios that may be expected in the system of interest. In this regard, the three factors below must be taken into consideration:
   a. The manufacturing standards and procedures for making the system. For instance, the welding or hydrotest procedures that may have been applied.
   b. Corrosion management measures, whether the system is under cathodic protection, whether coating has been applied etc., and whether they have been applied in accordance with the standards and codes. An example is that although welding may have been applied, but it has not been in accordance with the relevant codes.
   c. The working conditions dictated by the industry.
4. All possible corrosion scenarios (whether series or parallel corrosion scenarios) must be thought of and listed.

Although in next chapters we will explain more what we mean by "series" and "parallel" corrosion scenarios, it is essential to touch the basis here briefly about these two important concepts. It is because without having a sound understanding about the possible corrosion mechanism and/or scenarios in a corrosion system, it is too hard (if not impossible) to be able to define the possible root cause of the failure that may have resulted in a leaking "incident" that, because of its environmentally destructive nature, has been turned into an "accident."

Under natural, normal working conditions, it is very rare to face with corrosion cases in a given system where only one specific corrosion process is the main trouble-making issue. The point is that there are competing corrosion possibilities that because of many reasons one may appear to be the cardinal process. This will not exclude the possibility of other corrosion processes, and therefore we have to take care of knowing all possible corrosion processes that can be assumed to exist in the given system of interest.

Let us try to superimpose the four steps mentioned above about CSIs for a given system of interest. Suppose that the system of interest is a hydrant, that is to say, a fire water ring.[a] To evaluate corrosion of system impacts (CSIs) algorithm, we will follow the following steps:

1. What is the material of the ring? If it is a nonmetallic material such as composite materials, then the risk of corrosion is highly unlikely. However, if it is made of a metallic material such as carbon steel, then the risk of corrosion will exist. Although it is not an engineering practice to make these rings from stainless steel or titanium alloys, by making these materials the manufacturing material of the ring, the risk of corrosion, and MIC would be highly reduced.
2. If the ring has welded parts is also a question to be answered. Fire water rings neither are not short pipes nor are seamless. Therefore, they can always be classified as welded steels. They are not internally coated nor are subjected to hydrotesting. Therefore, the only standards that must be taken into consideration will be codes and standards related to welding and postwelding treatment.
3. a. Is the welding applied in the way it was supposed to be? Precommissioning inspections as well as inspections applied because of failures may prove to be good means to answer to this question. Bear in mind that from an ecologic-corrosion point of view, the main guideline for the operators will be "prevention" and then "control" of the pollution.
   b. Is the ring under any cathodic protection? Does it have any lining? Is the water treated with biocides based on a presurvey, the result of which have shown the CRB/CRA[b] that may have been present in the water? Is there a routine procedure by which parameters such as TDS (total dissolved solids) as well as the number and type of the CRB and CRA are being monitored?

[a] A more detailed corrosion management approach for a fire water ring is to be presented in Chapter 3 to explain "Javaherdashti corrosion management model."

[b] In the following chapters, we will explain what we mean by CRB (corrosion-related bacteria) and CRA (corrosion-related Archaea).

c. The working conditions of the fire water are that it is most of the time stagnant and at ambient temperatures. The flow rate of the water is less than 1.5 m/s most of the time.

4. There can exist several corrosion scenarios that in essence could be under deposit corrosion caused by the differential aeration cells, galvanic corrosion between the parent material and the weld, and MIC. Of these corrosion scenarios, MIC will be the most important one that can contribute to underdeposit corrosion as a series of corrosion scenario (by producing more biofilm, more anaerobic/aerobic spots may be created, leading into intensified underdeposit corrosion. With the same token, MIC can act as a parallel corrosion scenario to the galvanic corrosion that may be expected: the anode/cathode thus produced by inappropriate materials and within the conductive water that will be acting as an electrolyte will continue to affect the seriousness of corrosion; meanwhile, MIC is proceeding with its own mechanisms).

As seen from the above, the CSIs of the fire water ring can be evaluated in this way. The countermeasures that can be applied to control corrosion could themselves have environmental impacts, thus giving rise to CCIs (= countercorrosion impacts). For example, even seeming "green" countercorrosion practices such as cathodic protection could have detrimental impacts on environment.[7]

Fig. 1.1 schematically describes the interrelation ship between CSJs and CCIs[2]:

Fig. 1.2 shows discharge of hydrotest water[8] as reported to one of us (Javaherdashti) into the open seas. The water with a darker color that is being discharged from through a pipe contains CRB as well as toxic gases such as hydrogen sulfide. The water had been left as a result of hydrotest in the pipe for some months. The dark water and why it had been formed is a subject of "corrosion effects" because to understand why the seawater that had been used as the hydrotest water had become so corrosive, one needs to know the main corrosion scenarios that govern posthydrotest corrosion and also that why MIC is the prevailing one. On the other hand, as this water that enters into the sea does have contaminating agents in it, the water is also a case of CSI: this contamination in the discharged water could have detrimental impacts on the ecology of the sea, even if it is local. This means that because of the corrosion thus generated, an impact has

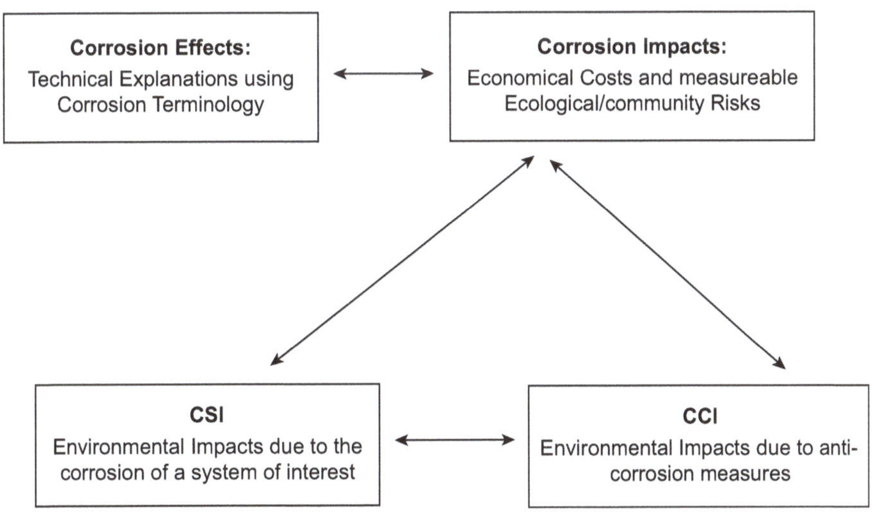

FIG. 1.1 How CSI and CCI are connected to expand the meaning of corrosion impact.

FIG. 1.2 Discharge of hydrotest water (dark water from the pipe) into the open sea (collection of Dr. Reza Javaherdashti).

been created that has the potential of inducing risks of ecologic nature to the marine ecosystem. However, the hydrotest water also contained a cocktail of chemicals (corrosion inhibitors, biocides, oxygen scavengers) whose environmental impacts were not surveyed and studied during the time Dr. Javaherdashti was working as a corrosion consultant to this project. This means that the hydrotest water could also have the capability for being called as CCI, as we don't know about how the chemical cocktail could have had adverse impacts on the ecology.

The example shown in Fig. 1.2 can clearly show that a case could have various aspects when looked at from different viewpoints, and therefore corrosion and environment could be tightly entangled issues that it could not be easy to separate them from each other. Another example of possible ecologic impacts of corrosion has been given elsewhere.[9] Javaherdashti[6] has reported cases where MIC has been identified as the main cause of the catastrophe. However, we want to once again emphasize on the fact that evaluation of ecologic aspects of corrosion especially MIC is a multidisciplinary task and it requires a well-documented approach. Just reporting that such and such disaster happened because of MIC is not making that use. What makes it count is to define the cause and effect and try to concentrate on corrosion effects as well as corrosion impacts in these cases.

How can Risks, Defined as a Function of Likelihood and Consequences, be Related to Microbial Corrosion?

A very common source of industrial managers not being able to see that MIC disaster coming is to mix up "likelihood of MIC" with "risk of MIC". The main reason for this, our years of experience with various industries around the world, is that although risk is a well-known concept in HSE (health, safety, and environment), it is not a familiar matter to corrosion professionals so much so that they have not been able to properly communicate this with their managers. Although we accept that there can be exceptions to the rule and many of these exceptions may indeed be too evident to deny, we can see it from the never-ending list of incidents that could otherwise be simply prevented or to a large extent controlled, similar to those reported.[10] To start with, it will be good if we look at the main structure of a risk matrix (Fig. 1.3):

There are three elements in a risk matrix that must be taken into consideration:

a. The element of likelihood
b. The element of consequences
c. The element of risk classification

"Likelihood" is the main task of a corrosion engineer: he has to assess if a corrosion (and for that matter, an MIC) hazard is nearby, and if yes, how close it is. Normally, one way of saying "how close

Risk assessment matrix						
		Consequence				
		Low	Minor	Moderate	Major	Critical
Likelihood	Almost certain	High (11)	High (16)	Extreme (20)	Extreme (23)	Extreme (25)
	Likely	Moderate (7)	High (12)	High (17)	Extreme (21)	Extreme (24)
	Possible	Low (4)	Moderate (8)	High (13)	Extreme (18)	Extreme (22)
	Unlikely	Low (2)	Low (5)	Moderate (9)	High (14)	Extreme (19)
	Rare	Low (1)	Low (3)	Moderate (6)	High (10)	High (15)

FIG. 1.3 How a risk matrix may look like.

the corrosion danger is" will be by applying corrosion assessment and monitoring techniques. One way of doing it is, that is, by placing a metallic coupon of the same material of the system of interest in the actual working conditions and observing that how much of its initial weight will be lost because of corrosion in a given time length. This method is addressed as "weight loss corrosion rate assessment". This is a mimic of the actual corrosion scenarios in action and reports the output as a function of mass loss and time, a measure that is reported as "corrosion rate." To demonstrate how corrosion rate can be a useful tool in this context, we can think of a pipeline whose thickness is, say, 2 mm. If weight loss measurements came up with a corrosion rate of 1 mm/year, this means that, on the average, it can be expected to lose 50% of the thickness of the pipe in 1 year and provided that no rehabilitating measures are provided, the whole pipe within 2 years of operation.

Although these techniques (and other techniques that are widely available in corrosion engineering to mimic corrosion and give estimations about the so-called "remaining life" of the structure") are all in use in industries nowadays, we have to be very cautious about their limitations too. For instance, the assumption behind using coupons in weight loss studies is that the corrosive environment will be corroding the material in a uniform manner (uniform corrosion) rather than accelerating corrosion at one spot (localized corrosion). We will briefly explain uniform and localized corrosion in the next chapter but for the time being it is sufficient to just say that the assumption of existing a "uniformly corroding environment" is not necessarily true everywhere and in most cases of corrosion, and particularly in the cases of MIC, the majority of corrosion morphology resembles localized corrosion in the form of local miniscule holes (i.e., pitting) that will later coalescent to form larger pits or even cracks and by lowering the effective surface to carry the load, the mechanical stability of the material will decline, leading into failure. The data from weight loss study are not of any use as individual data, it is the trend they make that will give us the information we will need to make a judgment about how a system is responding to the threat of corrosion. In other words, if the collection of data points will make an ever-increasing trend, that may be the start of the end. Fig. 1.4 will schematically show the possible situations that may exist in an imaginary system of interest for monitoring corrosion in it:

Although the likelihood can be measured quantitatively in the literature of corrosion as we mentioned above with the example of weigh loss method, the consequences may be interpreted as of limited knowledge to a field, corrosion engineer. In fact, what can be seen in practice is that the corrosion engineer can estimate the immediate outcome of a potential corrosion hazard such as leakage from a pipe that may cause production delay or even explosion. The assessment of the aftermath issues, that is, the extent of CSIs, often remains uncategorized and nonquantified to a corrosion engineer working in a

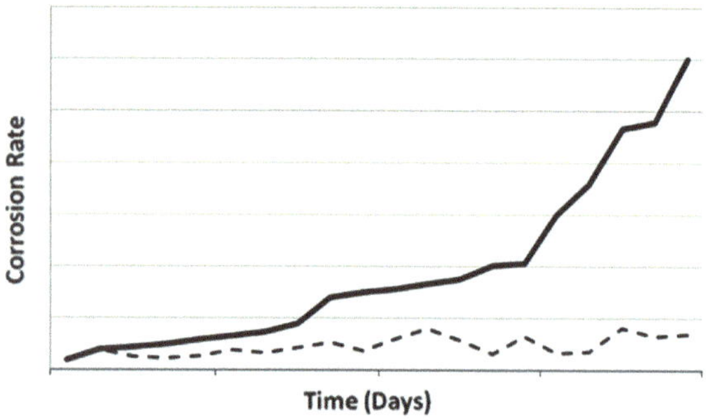

FIG. 1.4 Two imaginary corrosion rates vs. time for a weightless assessment pattern: the dashed line shows that during the test duration, corrosion is almost predictable within a certain fluctuating margin; however, the bold line shows that the corrosion rate is ever increasing. What is to be seen here is that in both cases, it is not the absolute values of corrosion rate but the trend they make that is of importance.

refinery or a power plant. Not that this particular individual is not concerned about the possible impacts on the environment that a corrosion effect may cause, but that simply it has not been defined for him.

We don't want to spend too much time on the consequences of corrosion (and for that matter, MIC) here, but we have to advise that without a clear understanding of these consequences (both CSI and CCI impacts), although a corrosion has done his job as a corrosion engineer, as a member of the greater society of people living in a community, he has an unfinished business. The readers of this book should not think that in the developed countries, all measures against CSI and CCI are taken. This is largely true but not the whole picture. We are mainly concerned about the decision-making minds and positions that are only busy with keeping their production high no matter the cost (Fig. 1.5).

The risk, as we mentioned earlier, is how the likelihood and the consequences factors interconnect and act. In other words, a high risk may be a product of an unlikely likelihood and a very major consequence. An in-depth survey of the causes of many MIC-related disasters can show us that the risk and the likelihood had been mixed up and that resulted in the problem. This mix-up can take various shapes, for example, just because biocides had been applied or just because corrosion monitoring has not resulted in any particular "sudden change," it is assumed that corrosion in the form of MIC does not exist. Here we want to explain a firsthand experience that one of us (Javaherdashti) had in one of his site visits to a

FIG. 1.5 We don't want such mentalities and mind-sets. We are sure that not only within this company but also other internationally reputable companies there are strict rules and policies to take care of our environment, but considering that a pill could be the direct result of corrosion (corrosion effect) and could have various, detrimental impacts (CSI), it is necessary to consider corrosion and its risk very seriously. (Source: http://www.vancouversun.com/.../Kinder+Mor.../9793673/story.html.)

very reputable petrochemical complex: when we were called to this complex, the operators were only concerned about the leakages they were getting in their fire water (hydrant) systems. As a matter of fact, they did have such leakages and we did observe them. We also noticed that the complex was using wet cooling towers to cool off the steam later to be used for electricity generation. We asked if they have had any MIC problem with this water and their answer was negative. When we insisted on understanding why they are of this opinion, they replied that they had an MIC monitoring system in place for monitoring the likelihood of MIC in their cooling towers. The monitoring system was in essence consisting of a series of glass loops within which there were carbon steel coupons. The loops were filled with the water that was coming from the towers, and the assumption was that if the water is contaminated with the CRB, then biofilm must be observed on the coupons. The assumption did sound right though. The issue was that they had not been observing any biofilm of considerable size formed. Therefore, they were assuming that no seeing biofilm on their coupons was a good reason to assume that no MIC threat (risk) of considerable value was present. We also have to mention that although the company that had made and installed the monitoring system had advised the operators to remove the coupons every 45 days, because of production schedules, the production management was only allowing them to do the monitoring period for 3 weeks (!) and not 45 days. When these operators tool an MIC workshop, they realized a simple yet detrimental error in their MIC monitoring procedure: they were installing the coupons on a vertical position instead of a horizontal position! By a vertically positioned coupon, the gravitation force will be one of the major factors not allowing any deposit, let alone biofilm, form. As seen here, the assessment of the risk based on wrong likelihood calculations could have very serious issues and aftermath none of which was taken seriously by the top management until a serious problem arises.

In the next chapter, we will talk about corrosion as much as required to understand the mechanisms of MIC.

REFERENCES

1. Hou B, Li X, Ma X, et al. *The Cost of Corrosion in China*. www.nature.com/npjmatdeg. Published on-line, 25 July 2017.
2. Javaherdashti R, Nikraz H. *A Global Warning on Corrosions and Environment: A New Look at Existing Technical and Managerial Strategies and Tactics*. Germany: VDM; 2010.
3. Koch G, Varney J, Thompson N, Moghissi O, Gould M, Payer J. *International Measures of Prevention, Application, and Economics of Corrosion Technologies Studies (IMPACT)*. USA: NACE International; March 2016.
4. Flemming H-C. In: Heitz E, Flemming H-C, Sand W, eds. *Economical and Technical Overview in Microbially Influenced Corrosion of Materials*. Berlin, Heidelberg, Germany: Springer-Verlag; 1996.
5. Parthipan P, Narenkumar J, Elumalai P, et al. Neem extract as a green inhibitor for microbiologically influenced corrosion of carbon steel API 5LX in a hypersaline environments. *J Mol Liq*. 2017;240: 121–127.
6. Javaherdashti R, Nwaoha C, Tan H. *Corrosion and Materials in Oil and Gas Industries*. USA: CRC Press/Taylor&Francis; 2013.
7. Rousseau C, Baraud F, Leleyter L, Gil O. Cathodic protection by zinc sacrificial anodes: impact on marine sediment metallic contamination. *J Hazard Mater*. 2009;167:953–958.
8. Javaherdashti R, Akvan F. *Hydrostatic Testing, Corrosion, and Microbiologically Influenced Corrosion a Field Manual for Control and Prevention*. USA: CRC Press; 2017.
9. https://www.linkedin.com/pulse/note-link-between-corrosion-management-environment-reza-javaherdashti.
10. Kletz T. *What Went Wrong? Case Histories of Process Plant Disasters*. 4th ed. Gulf Professional Publishing; 1999.

CHAPTER 2

A Brief Introduction to Corrosion Engineering

INTRODUCTION

In this chapter, we will briefly introduce the fundamental concepts that are necessary to understand electrochemical corrosion, which is the basis of any corrosion studies required for studying MIC.

THE THERMODYNAMICS OF CORROSION AND ITS IMPORTANCE IN DEALING WITH IT IN INDUSTRIES

The very word "corrosion" comes from a Latin word that originally means "bitten away." In many other languages, the terminology used to address corrosion is either a "dialect alteration" of the English word (e.g., corrosion in Spanish) or even if equivalents are used (e.g., in Persian خوردگی or in Arabic تاكل), they both would bring the same concept to their readers: bitten away.

There are a few confusions about corrosion. These confusions do not just come from the electrochemistry involved but even in daily addressing of the terminology these confusions arise. Two of very famous of these confusions are about rusting and corrosion of nonmetals.

Many people—even professional engineers—mistake corrosion with rusting: in fact, all types of rusting are corrosion but not all types of corrosion are rusting. In more precise terms, if ferrous alloys (for instance, steels) undergo corrosion, we can call it an example of rusting. It is more accurate if we say copper or aluminum corrodes and not rusts.

The second confusion in addressing nonmetals and their so-called "corrosion" is a little bit more confusing. Nonmetals such as polymers, concrete, glass, composites, and the like cannot corrode. Or better to say, they cannot corrode electrochemically. As we will see in the next sections of this chapter, electrochemical corrosion does need some underlying elements to be present for it to happen. One of these elements is having a conductive path for the flow of electrons. None of the nonmetals we named above do have such criteria. In fact, because of their very electronic structure, no electron flow of the same we observe in metals can be thought of for nonmetals.

Therefore, nonmetals do not corrode simply because they cannot. They degrade and deteriorate that if we really want to distinguish it from electrochemical corrosion, deterioration and degradation are rather physical processes in which no electron/ion exchange can be expected.

Corrosion is important not just with regard to what we already mentioned in the "Introduction" chapter but in the sense that it is, to a large extent, "unavoidable" but "controllable." In this chapter, we will deal with two main questions: why is corrosion unavoidable and why it is controllable.

The story of corrosion starts with the way that civilization started to know the use of metals. Metals are all "hidden" in their ores. One has to find a way to take them out or, more scientifically speaking, extract them. But have we ever wondered why it is so?

Metals, like any other material, have been composed of atoms. These atoms have certain number of electrons that will make them, altogether, being iron, copper, silver, or any other metal. Under normal conditions, these atoms are all at minimum level of energy. In other words, they are at rest within their ores. That is why you will never see or expect metals to "ooze out" from their ores. We have to apply energy to extract

Biological Treatment of Microbial Corrosion. https://doi.org/10.1016/B978-0-12-816108-1.00002-1

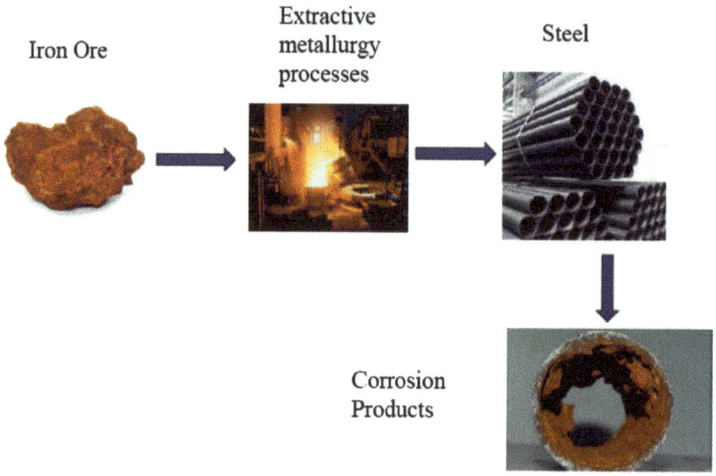

FIG. 2.1 The life cycle of iron from birth (extracted from its ore) to death (Corrosion).

them. When we do take them out from their "comfort zone," we are in fact putting more energy to them and make them unrest by injection of electrons during extractive metallurgy processes (Fig. 2.1).

The cycle, as shown in Fig. 2.1, is self-explanatory: the metal is extracted; during the extractive metallurgy processes; it receives extra electrons, and therefore, its excitation level, i.e., Gibbs free energy, is highly increased. The net result is that to retain to its original low energy level, the metal starts to, literally speaking, throw away these extra electrons. This process, otherwise being called corrosion, helps metal to reduce its excitation level and get back to its original low-excitation level. However, this will have its own toll on the metal too: the metal will not—from an engineering point of view—the original material but rather deposits that in essence are the same as what was originally embracing the metal in its ore: the metal will turn into red rust (iron hydroxide) or black rust (iron sulfide) or iron carbonate, etc. These all were the companions of the metal when it was still "not born" and was in its ore.

The importance of understanding this process has at least two advantages:

1. There is no way to stop corrosion. It is because corrosion is a thermodynamically favorable process. In other words, as long as we extract metals out of their ores, we have to also accept that corrosion is "unavoidable."
2. However, we can find ways to slow down corrosion or control it. In other words, we can provide the conditions by which the process of "electron throw out" gets slower. We can actually slow down corrosion by such magnitude that, from a practical point of view, it will get many years to survive. This is the reason we say corrosion is "controllable."

We would like to expand and explain the importance of the two points we mentioned above in more detail.

1. Regarding the point that corrosion cannot stop: we know from many encounters we have been having with industry that unless the client is well educated about the thermodynamics involved in corrosion, as soon as they hear that you cannot stop corrosion, they would take as a clear sign of your inability and noncompetence and would disqualify you as a corrosion consultant. It is comprehended that one cannot bend the truth for the sake of getting a job; however, the harsh reality out there— in many cases out, there is as we just pictured. Corrosion can be brought under control, but even this will not guarantee that without a continuous maintenance and monitoring, corrosion would be defeated. However, maintenance and inspection are not necessarily the same things: inspection must be an ongoing, continuous procedure, whereas maintenance is an intermittent action in time,

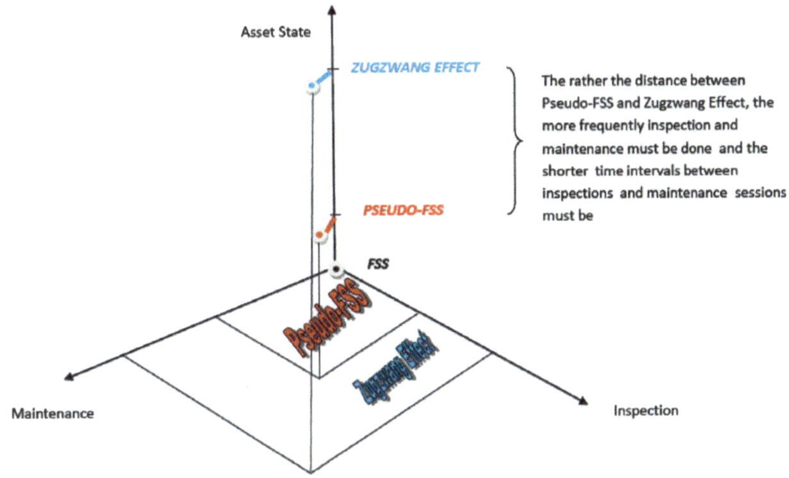

FIG. 2.2 Schematic representation between the asset's states (FSS, pseudo-FSS, and Zugzwang effect state) with maintenance and inspection. Black dot presents FSS, whereas red and blue dots present pseudo-FSS and Zugzwang effect states (not to scale).

equating it with repair. It is quite imaginable for a corrosion engineer to be engaged in inspection all the time but when the situation becomes critical gets involved in maintenance. If an asset is its ideal form of workability (workworthy or fit-for-service = FFS), it will not need any repairmen activity; however, it will still need inspection because as soon as it arrives into service, it will be exposed to wear and tear and corrosion. This deviation from FFS state, which we call pseudo-FFS, will still allow using the equipment, but this time the vulnerability of the asset will obviously increase. This will enhance the likelihood for maintenance that can be done at prearranged schedules, but yet inspection will keep its continuous need for applying. This procedure will continue until the equipment, if not being inspected and maintenance is not being properly done, will develop shutdown periods (nonscheduled overhauls) and will arrive at a state that we call as "Zugzwang effect" state: at this state, either it is economically not feasible to conduct repairs or even if repairs are being made, as long as the root cause of failures has not been recognized, overall well-being of the asset is always under question. These states and their relationship to inspection and maintenance can schematically be shown as in Fig. 2.2. Fig. 2.3 shows an example of an asset that has arrived at Zugzwang effect state:
Fig. 2.2 simply emphasizes upon the fact that the longer an equipment/asset remains in action/service; it will need more maintenance and more frequently inspection.

2. Slowing down corrosion: This is a very important issue because as long as we know no stop to corrosion can be achieved, the only alternative will be finding effective means to decelerate it. A very important issue here, however, is to distinguish between corrosion control, corrosion prevention, and corrosion stop.
   a. When an asset is put into service and is being used, we face with an issue that is already done and little can we do about it. In other words, when we are facing with corrosion in a plant that has been going for some years since its establishment, it means that we are dealing with equipment that have been designed and manufactured and the manuals that have been given to run those equipment. We cannot just pop in and say that to manage corrosion, it is necessary for the operators to change or modify the process parameters (design, process temperature, process pressure, process chemistry (for instance, the designed pH), materials). By changing

FIG. 2.3 An example of Zugzwang effect on a concrete wall. This wall was part of a cooling tower structure that was directly facing the salt spray from seaside. The concrete has been damaged very badly. In this figure, most probably two parallel corrosion processes have started the deterioration process: the salt spray has affected the integrity of the concrete, whereas the moisture ingress has caused corrosion of the steel bar. The steel bar has gone through rusting so that the corrosion products thus produced have mechanically damaged the concrete layer, therefore letting more moisture and salt ingress and consequently further deterioration. The asset is in Zugzwang effect state because any applied repair is doomed to fall off as long as no solution for the salt spray and moisture ingress has been thought about as a root cause solution. (From Dr. Reza Javaherdashti Collection.)

each (let alone all) of these process parameters, chances are that the whole process would be affected (probably adversely) just to manage corrosion. No manager would buy that. However, under these preestablished conditions, the only way by which corrosion can be managed will be controlling corrosion by applying the known procedures, standards, and the like. For instance, although the pH of the process is designated at, say 4—which based on corrosion potential may be leading to corrosive conditions (consult with the related Pourbaix diagram)—we can still use addition of corrosion inhibitors or applying specific coatings or even go for materials selection (provided that it will be justified) without trying to change the pH of the process. Corrosion control strategies will suit equipment and assets at plants, which have been running for some years, in other words, a brownfield project.

b. Corrosion prevention, however, is another story: it can be applied into a Greenfield project. When a plant is still at design stage and perhaps early stages of construction, it is possible to design and apply strategies with the general aim of preventing corrosion. Here what we mean by corrosion prevention is to reduce corrosion rates to such low values that practically they can be taken as almost-zero rates. It is at this stage that by applying proper plans and precautions, corrosion prevention may become possible. Examples of such precautions may be as follows: performing hydrotest in such a way that will not be leading into posthydrotest corrosion (MIC), conducting preferring OPEX (operation cost) over CAPEX (Capital expenses) when it comes to corrosion prevention, giving care not to have too many dead legs and zones in pipelines, emphasizing upon flow speed rate rather than flow volume rate (in other words, to have meter per second values for the fluid in parallel to having liters per second), and the like. Corrosion prevention is the main step in dealing with corrosion from a practical point of view for making corrosion rates almost-zero. For corrosion control, corrosion rates are to be always lower than certain KPIs set by standards and regulations, whereas for corrosion prevention, those standard values are almost automatically achieved.

c. Corrosion stop is an imaginary, ideal state of the service life of the asset in which the corrosion rates are indeed zero. As we mentioned earlier about the thermodynamics involved in corrosion, achieving corrosion stop is thermodynamically impossible and therefore it can only be taken as a fantasy.

In the following section, we will explain some technicalities related to corrosion with more details.

SOME BASIC CONCEPTS OF CORROSION

Our main focus here will be on electrochemical corrosion. The main reasons for focusing on electrochemical corrosion are as follows:

1. The main medium in which electrochemical corrosion takes place is water.
2. Water is also the essential medium for bacteria to live in and survive.

Therefore, if there is no water, one can safely assume that neither electrochemical corrosion nor microbial corrosion would bother anybody! The reality, however, is that as long as we are living on the earth, there is no way to exclude water. This water can either be an essential part of a process (e.g., water and waste water treatment) or production-support system (cooling systems) or entrained water that is normally found with fluids such as oil and gas and at low-lying points of pipelines will be collected to form small-size ponds.

Therefore, from now on, in this book, when we use the term "corrosion," we will mean "electrochemical corrosion" unless advised.

ESSENTIALITIES FOR CORROSION

Simply put, when metals of different characteristics (such as, but not limited to, their ease of release or gain electrons) are brought together in a watery medium, the metal higher tendency for losing electrons will start to corrode. This metal is scientifically and technically referred to as the "anode." Its counterpart at the other end of the system that will receive these electrons will be called the "cathode." This simple description should not confuse our readers that the electrons that leave the anode will "swim away" from it and find their rescue on the cathode. In fact, these electrons will need a metallic path to be able to reach the cathode, and this metallic path is the metal surface itself. The ions that are produced via anodic reactions (that is, the reactions that will give off electrons) and cathodic reactions (electron-gaining reactions) as a result of electron release and gain will be traveling through the water (that we will refer to it as electrolyte).

From the above, it may follow the following:

1. For corrosion to start, one has to have at least two metals, which are in contact via the electrolyte
2. These metals must be in such a way arranged to sit that will have different potentials (capabilities for losing and gaining electrons)
3. During corrosion process, the electrons lost by the anode will be traveling to the cathode via a metallic path, whereas the ions thus produced (positively charged ions = cations, and negatively charged ions = anions) will be traveling via the electrolyte toward cathode and anode, respectively.

The process is simply shown in Fig. 2.4.

However, from the above, we can also conclude the following:

a. Without an electrolyte or a metallic path, no corrosion is expected to happen.
b. Corrosion cannot take place on nonmetals.

Conclusion (a) is obvious: if we omit either electrolyte or the metallic path, the circuit will be incomplete. In other ways, there will be no travel path for electrons and ions. This will explain why it is essential to do "coating"—we will get back to it later in this chapter. Conclusion (b) is even more critical: it explains that plastics, composites, and the like do not corrode. Their failure mechanism, whatever it is and we will not focus on it here, cannot be referred to as corrosion simply because no electron exchange can be expected to take place even if there can exist an electrolyte.

Also, as the careful readers have noticed in Fig. 2.4, the way we have described the essentialities of corrosion differs somewhat from what that can be found in classical corrosion and electrochemistry books. In these books, four

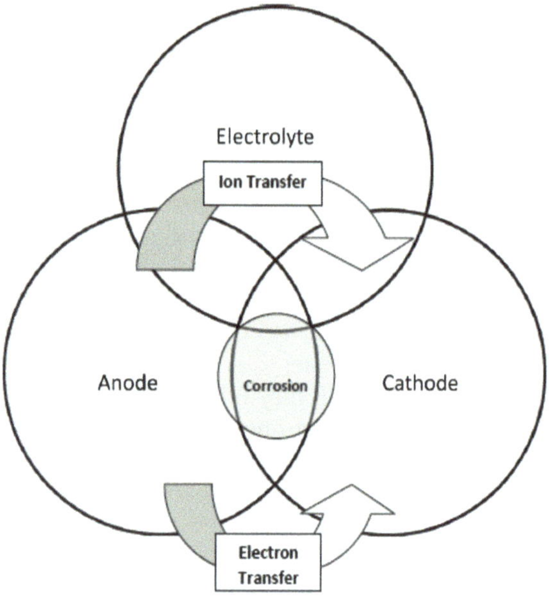

FIG. 2.4 Essentialities of corrosion in terms of electrodes (anode/cathode) and the electrolyte (the ion transfer medium).

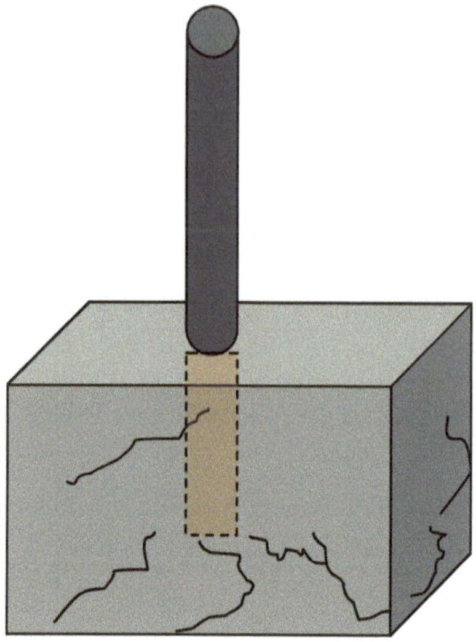

FIG. 2.5 An example of series corrosion processes.

elements are described to sustain a corrosion process: anode, cathode, electrolyte, and a metallic path. We have omitted the condition for a metallic path as in our understanding of real life conditions under which corrosion occurs that it does occur when a metallic path is available. Thus, we see no need to reemphasize upon an evident fact!

"SERIES" AND "PARALLEL" CORROSION PROCESS PATTERNS

We may remember from our high school physics about how we can have two combinations of electric circuits and how they would affect the total output current as well as the resistance. Those combinations were "parallel" and "series" combinations. We can use those in our discussion too.

It is possible that sometimes two or more corrosion processes happen to occur in such a way that they may enhance each other. This is what we refer to as "series corrosion processes." An example is as shown in Fig. 2.5:

A COMBINATION OF A CONCRETE SLAB AND A STEEL ROD

Assume that a steel rod has been placed inside a concrete slab. The concrete is not healthy in the sense that it has developed several pores and cracks of different sizes and orientations. In this example, if water ingress takes place through the pores and cracks on the slab, this will reach into the inner section of the rod, thus rusting it. Obviously, the more the water ingress takes place, the higher the likelihood for the inner part of the rod inside the slab to corrode. Why do we call it an example of a series corrosion process? The rusting of the rod is being enhanced by the water ingress through the concrete pores. In other words, if deterioration of concrete that has resulted in the formation of the cracks is stopped (so that no pores and cracks are developed), then the rusting of the inner section of the rod will also be stopped. A practical use of formulating the situation as such is that instead of focusing on the rusting of the rod per se and trying to find a solution for it, it is advised to find a solution to control the pores in the concrete slab via mitigation of the concrete deterioration.

Now assume that we have done no special treatment on the slab yet so that it still developing pores and cracks in it. However, we have coated the inner section of the rod with a material, which is resistant to corrosion; for example, we have wrapped a polymer around it so that it will not allow the water that has been engrossed via the pores and cracks on the slab surface to reach on the surface of the steel rod inside the slab. Obviously, this will solve the problem of corrosion of the rod inside the slab. Yet, there is another scenario, which is going on in "parallel": the upper section of the rod is still exposed to atmosphere and this will make it even easier for to corrode. In this situation, the corrosion of the outer section of the rod and the deterioration of the concrete slab are two corrosion/deterioration processes that do not depend on each other, and therefore, they happen independently, an example of "parallel corrosion processes." The practical moral of the story is that if you cure and fix the deterioration problem of the concrete slab, you cannot expect it to have any effect of the corrosion of the outer section of the rod and vice versa.

Understanding different corrosion scenarios as such could be of great help: the possible galvanic interaction between the iron sulfide film produced by sulfate-reducing bacteria and overall MIC thus occurring can be taken as an example of series corrosion processes (the more the FeS produced by SRB, the more intense the galvanic corrosion effect). Another example could be the link between the number of acid-producing bacteria and the overall corrosion rate: the higher the number of APB, the higher the expected corrosion rates. An example of parallel corrosion processes with MIC is that in the same petrochemical complex, the fire water ring and cooling tower of the base load power plant of the complex both corrode via MIC but have no dependence on each other.

The immediate results for treatment strategies as in given examples above are stop MIC to stop the galvanic effect and in the petrochemical complex spend budget on the management of MIC in both the units separately. Understanding corrosion as such will decrease a lot of unnecessary expenditures of financial as well as man power resources.

HOW IS CORROSION TREATED?

Naturally, a big concern for those industries involved in dealing with corrosion is how to reduce its detrimental effects. Over the years, many strategies and tactics have been developed that can be generally categorized as electrical measures, physical barriers measures, chemical measures, as well as materials and design measures. In other words, in fact, it is possible to categorize these measures as the main methods by which corrosion can be controlled (remember, corrosion cannot be stopped because of its thermodynamical nature; it can only be controlled):

1. Application of cathodic protection (CP) and to a lesser extent anodic protection (AP) methods
2. Materials Selection
3. Application of coatings
4. Design factor
5. Chemical treatment

Although CP and AP are electrical measures, application of coatings can be regarded as creating physical measures against corrosion (removing contact between electrolyte and anode/cathode as well as physically separating anode and cathode from each other so that anode and cathode will not "see" each other). Likewise, application of chemicals such as corrosion inhibitors, biocides, and the like is an example of chemical treatment. We will discuss materials selection and design factor in required length in the following sections.

There are two points here that are worth of being considered: the first point is that based on the complexity of the corrosion system (the system in which corrosion is occurring), it may be required to use all of the abovementioned five corrosion management measures. For instance, for a buried pipeline, to cover for its external corrosion (that occurs due to its exposure to the soil environment), both CP and coating will be applied. To manage the issues of internal corrosion, chemical treatment application is a must. Also, based on the severity of corrosion and cost justifications, it may be possible to change the material of the pipe (this could, for example, be switching from carbon steel to composite or polymeric materials). Design factor is also very important in the sense that potential

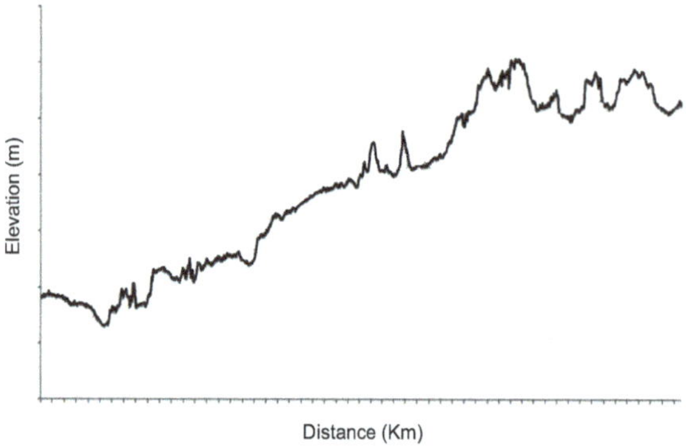

FIG. 2.6 A typical (subsea) pipeline topography scheme where elevation is recorded as a function of distance from a given fixed point (e.g., a pumping house). (Photo: from Dr. R. Javaherdashti collection.)

for microbial corrosion will increase if the pipe happens to have "too many" branches such that likelihood of fluid stagnation would increase. Yet another factor that may be related to increasing the likelihood of corrosion (in both microbial and nonmicrobial forms) is the topography of pipeline laying (Fig. 2.6):

As it can be seen in Fig. 2.6, pipelines in reality are not laid up on straight lines: they pass over hills and have their own topographical ups and downs. Although it may not seem directly linked with what we may have in mind from design, topographical situation of a given pipeline can intensively increase the likelihood of corrosion (especially in the form of internal corrosion). In Fig. 2.6, low points of the topographical map where, owing to low elevations, the line will face points where fluid stagnation becomes quite likely are also the points at which the risk of both microbial corrosion and nonmicrobial corrosion (that is, purely electrochemical corrosion that does not involve direct or indirect involvement of corrosion-related bacteria and archaea) will highly increase.

We would like to once again remind the reader that the main topic of this book is corrosion as related to living organism, that is to say microbial corrosion, and one of the ways by which microbial corrosion can be addressed is by application of certain chemicals, including biological biocides (to be discussed briefly later in this chapter). Therefore, what we will discuss here is a short introduction to each of the five factors above without trying to go through the details. The reason is, obviously, that our readers should expect this chapter and the next chapter (microbial corrosion) to be only taken as introductions to corrosion and microbial corrosions precise enough to understand the significance of why biological biocidal treatment of microbial corrosion is necessary.

APPLICATION OF CATHODIC PROTECTION METHODS

To understand CP, it is necessary to get back to the basics of corrosion, anodic and cathodic reactions. We said earlier that it is via losing electrons by anodic reactions and receiving them by cathodic reactions that corrosion current is created. In other words, if electrons cannot be lost or, alternatively, can be fed back to the reaction, then theoretically no corrosion can be expected. The first possibility, that is, not losing electrons, is impossible. The reason is that because of the thermodynamics of corrosion, metals in their fabricated forms, which, by the way, is how we use them in our everyday lives, cannot stop losing electrons. However, the other way round is possible: if we feed the lost electrons back to their ions so that they can again become atoms, then no corrosion will happen in essence. This "giving back the electrons" is the basic rule applied in CP practices. The electron give-back can be done via two methods: injecting electrons from an outside source (for example, a rectifier) or letting another metal, which in comparison with our main structures to be protected is of less value, corrode faster. In the latter case, what happens is that by letting the other metal corrode, an extra source of electrons is produced so that the required pool of electrons, which will be necessary to let the "electron give-back process" complete, will be provided (Fig. 2.7).

Fig. 2.8 shows an example of CP by sacrificial anode in practice; in this figure, the same steel bar has been put into the same corrosive environments (salt water simulating sea water) at the same time. The main difference is that in the left-hand side jar, there is a sacrificial anode attached to the rebar, whereas the right-hand

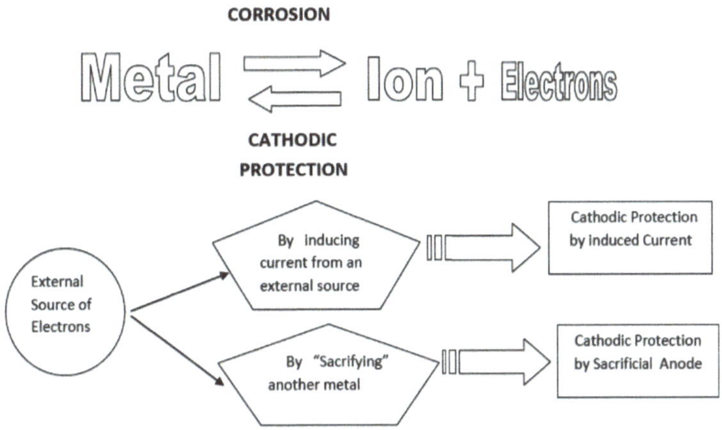

FIG. 2.7 A schematic summary of what cathodic protection in essence is and its different two methods of application.

FIG. 2.8 The story of two jars!, the left-hand side with CP and the right-hand side without CP. Except sacrificial anode CP, all other conditions for the two jars are identical.

side jar does not have it. As clearly seen from the figure, the jar with CP has remained pristine, whereas the one without it has already corroded (=rusted) and therefore will soon be losing its mechanical integrity.

MATERIALS SELECTION

What is actually meant by material selection is in fact the measures by which a material (alloy) is preferred over another one. After metals are extracted from their "nonmetallic shell," they are subjected to further (physical) metallurgical processes. In these processes, it is tried to induce various chemical and mechanical properties into the same metal to make several types of alloys out of it. For example, by increasing the carbon content of "iron," you will come up with cast iron, whereas by decreasing its content, you will get steel, also known as plain carbon steel. Now if we add up "some" chromium and molybdenum into it, you will get the so-called stainless steel. By increasing chromium content, you will get different classes of stainless steels, such as stainless steel 304, stainless steel 316, stainless steel 2205, and so on. But the "art" does not stop here! You can reduce the carbon content in stainless steel 316 and this time you will get 316L (L designated that it has low carbon). The possibilities could be endless. However, in addition to the physical, mechanical, and chemical properties of a material that can make it a good choice, there is also the importance of the cost (CAPEX=capital expenses). A certain material (e.g., titanium) can certainly have superior properties materials-wise over any type of stainless steel; however, costwise titanium is a very expensive material. Thus unless we have to, we prefer not to use titanium. This is a very important factor in the selection of thee right material. It is so important that a good number of corrosion failures happen because of wrong materials selection just because the material to be selected was simply "more expensive" and therefore the contractor would choose the less expensive option.

Another example could also be the wrong combination of right materials. In other words, while the materials selected for the intended service could be suitable regarding their corrosion resistance, their combination could be disastrous: a major petrochemical operator in one of the major players of petrochemical products countries in the Middle East experienced a bitter corrosion-related catastrophe in their plant in August 2010 that, in addition to the plant being out of service for at least 50 days, also cost the lives of five of the personnel.

In addition to careless operation, not observing standards, and other factors, the pipe whose rupture resulted in the disaster was in fact a big "battery" sold to this petrochemical unit. The pipe itself was made from stainless steel, whereas the rings around it to keep it secure were all carbon steel. This combination of dissimilar metals creates a dangerous combination known as "galvanic corrosion."

The pipe also had a corrosion-enhancing mechanism in addition to having another corrosion mechanism that was operative from inside: the fluid inside had been flowing with such a wrong angle of impact to the internal wall of the pipe that was actually acting as a hammer, eroding away bits and pieces of the pipe with itself.

APPLICATION OF COATINGS

Getting back again to the basics of electrochemical corrosion, as seen from Fig. 2.2, the main three factors in sustaining any electrochemical corrosion process are anode, cathode, and electrolyte. Therefore, if by any means we can prevent these three—or at least one of them—from "seeing" each other, then corrosion is expected to be controlled/stopped. This is the main feature of coatings: they prevent the electrolyte from seeing either electrodes (i.e., anode and cathode) as well as each electrode from seeing the other one.

For coating to act as required, it must be intact and must also have no scratches or holes (holiday-free coatings). This is important because otherwise corrosion will be enhanced at the holidays and discontinuities. This can be schematically represented in Fig. 2.9.

As seen from Fig. 2.9, when the material has coating, it has the least corrosion rate (the speed at which corrosion happens is the minimum). When it has no coating, it goes through corrosion but by time, the corrosion products are accumulated and they themselves act as coating, thus reducing corrosion. However, if it has a damaged coating, then it will experience the highest corrosion rates although again because of the accumulation effect of corrosion products, the corrosion rate will drop eventually. These laboratory findings clearly show that if it were in a natural system where the corrosion products could have been washed off from the surface and therefore new, fresh surfaces could have been exposed to the corrosive environment, then much higher corrosion rates than either coated or uncoated surfaces could have been expected from a damaged coated material.

Our reader must be reminded, however early yet, that the accumulation of bacterial cells, otherwise known as biofilm, can actually act like a damaged coating: this is why they are entitled to accelerate corrosion in most cases.

FIG. 2.9 (Left) Corrosion rates of coated, uncoated, and coating with damage of carbon steel coupons in simulated seawater. (Right) The damaged paint on a metallic bridge column, progressive corrosion due to damaged coating (paint), has severed corrosion. (Both figures from Dr. R.Javaherdashti Collection.)

DESIGN FACTOR

Design factor is the only anticorrosion measure that, instead of using a technology (electrical for CP, metallurgical for material selection, chemical for coating and chemical treatment), uses no particular technology. It is relying on the fact that certain geometries and/or practices can enhance corrosion, and thus by applying a simple change, one can easily get rid of the costs and risks involved. Reminding our readers of galvanic corrosion that, among other factors, can also happen because of dissimilar metals being coupled together, we want to give a case history here. In a power plant, the valve on a stainless steel pipe had been made of carbon steel. The dissimilar metals also had the small anode/large cathode ratio combination, leading into the valve being extensively corroded and being replaced almost every 2 months or so. The design from the beginning was not right, so to fix the problem what engineers did was to replace the carbon steel valve with a composite one. By eliminating the anode, no corrosion ever happened again. Figures below also show some examples of design factor in accelerating or increasing the likelihood of corrosion. It is a very important factor in controlling corrosion from the start. In fact, if there is any measure that can be designated as "corrosion preventive," it is the design factor.

Fig. 2.10 shows two examples where the design factor has been neglected. Most of the time, it is this negligence that can result in very serious corrosion and leakage issues. In industry, the important role of design and its significant role in the prevention of corrosion is not taken as serious as it should. In the upper picture of Fig. 2.10, it can be seen that the applied paint has been so poorly applied that it has added more corrosion issues to the pipe, which is already being corroded at the place of pipe supports. The picture below can easily pose a risk should any issue beyond expectation happens; for example,

FIG. 2.10 (Above) In addition to poor quality of the paint, corrosion has started from where the pipe is sitting on its support. With a simple design of the support, this corrosion could have been prevented. (Below) The welds are not continuous; therefore, a nonstandard spot welding has been applied. Corrosion has already started at one of the spots.

if the internal pressure of the vessel shown in the picture suddenly increases, the spot welds may not be expected to resist in a feasible way. Also, it is seen in the picture that signs of corrosion have already started at one of the spot welds.

A rather "worrying" aspect of design factor may appear when corrosion control becomes a requirement: in an already-established plant, it is almost impossible to think of adding or removing a process parameter such as design. This is despite the fact that, in practice, many industries do apply rather small modifications to their already-established assets. These small modifications do not count as they are not meant to change the overall design. In Fig. 2.2 (the below figure), changing welding into continuous welding does not affect the overall design of the part that is being used in the process. Those design factors would be faced with resistance from operators that would change the design thoroughly.

CHEMICAL TREATMENT

Chemical treatment of corrosion is a very widespread way of treatment. In fact, whereas in oil and gas industries it is quite common to see corrosion control measures such as CP, coating, materials selection as well as chemical treatment, in thermal power plants, for example, it is chemical treatment that can be taken as the most frequent way of treatment of corrosion (CP is not used in main power plants equipment). In addition, chemical treatment is one of the corrosion treatment options that has a considerable market size worldwide. Not only is this market including oil and gas, but also power plants, water treatment plants, and the like. The relative size of the world corrosion inhibitor market is seen in Fig. 2.11. This figure clearly shows that the main players such as North America, China, and Western Europe are the main sectors in which these chemicals are applied.

Corrosion inhibitors are those chemicals that affect anodic, cathodic, or both types of reactions in such a way that the overall corrosion rate is lowered considerably. This in turn would affect the severity of corrosion should either or both reactions control the rate of corrosion. In addition to inhibitors, there are other chemicals that can also serve to control corrosion. These have been schematically shown in Fig. 2.12. Just briefly

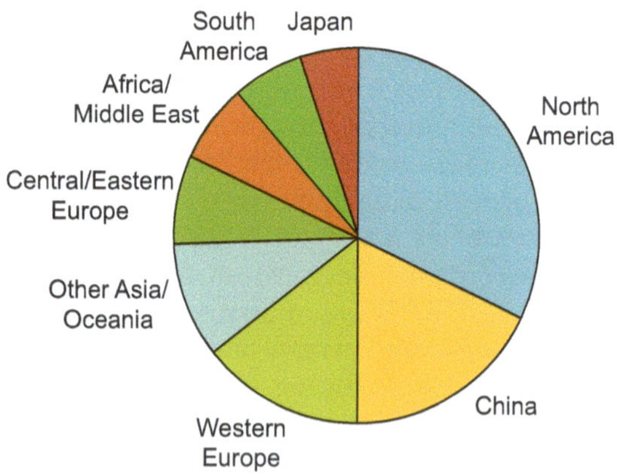

FIG. 2.11 World consumption of corrosion inhibitors in 2008. (After P.Rajeev, A.O. Surendranathan, Ch.S.N. Murthy-2012, used with permission.)

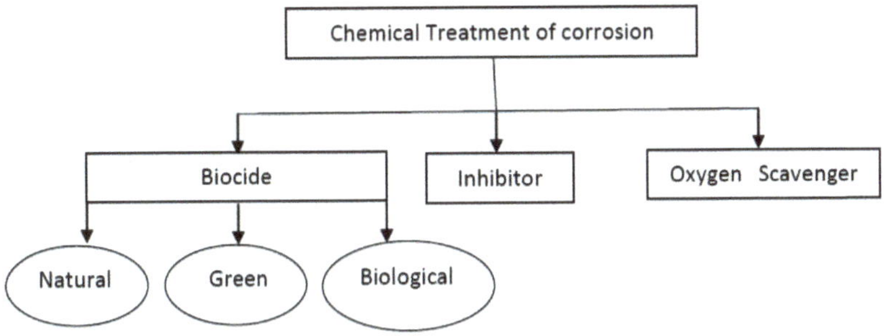

FIG. 2.12 Classification of chemical treatment methods applicable to corrosion cases.

said, we would like to say that the way oxygen scavengers serve to control corrosion is by excluding it from the system. Let us not forget that oxygen will act as a cathodic reaction facilitator so that if it is removed, even if electrons are liberated, there will be no reception site for them. Thus, the whole issue of corrosion will collapse. However, as long as microbial corrosion is concerned, it is the chemicals collectively known as "biocide"—literally meaning, killer of the living thing—that can help controlling the severity of microbial corrosion. As seen from the figure, in a rough yet right-to-the-point way, biocides can be divided into three categories: the so-called "green" biocides that are environmentally friendly and made up of formulations that are compatible with the environment, then the biocides to which we refer as "natural," and we mean any herbs, trees and their branches, flowers, roots, leaves, etc. that can kill microorganisms, such as bacteria. A good example of natural biocides is neem tree whose antibacterial properties have been proven by many researches around the world. In our judgment, all natural biocides are also green biocides but not vice versa. In other words, we believe that natural biocides (and also corrosion inhibitors, although we have not shown in the figure) are compatible with the environment per se obviously because they are—or once they have been—a part of the nature themselves. We may choose the essence or extract of such and such leaf and use it as a biocide in, say, the process of combating against microbial corrosion in a pipeline. The deal is that if for any reason there is a leakage of the fluid from within the pipe outside, then the biocide leaked into the nearby ecology has actually been itself a part of the natural environment once. This would mean that in the event of leakage, much less harmful chemicals than synthetic biocides would have contaminated into the environment.

As per the main topic of this book, we will study biocides in detail. In the second section of this book, biological biocides (bacteria against bacteria) will be discussed. It is, however, very important to note that to understand why biocides are useful, we have to know for what mechanism of corrosion they are applicable. This means that without a knowledge about microbial corrosion, however brief but right-to-the-point), it will be useless discussing biocides.

In the next section, we will introduce into microbial corrosion so much that is needed to justify why biocides would be needed and, even more importantly, why biological biocides—seemingly an oxymoron—must be studied in much more detail so that they become as commercially available and widespread applicable as their synthetic counterpart do.

CLOSING REMARKS

In this chapter, we showed that although corrosion is indeed a serious business, it is still curable should it is recognized and treated on time and monitored constantly. Corrosion cannot be avoided but controlled, this is a very important point that we will get back to it with more details when we are discussing the "**Zugzwang effect**."

A significant piece of corrosion management is to define the problem correctly and in its correct context. This means that without a good recognition, it is impossible to find the right treatment. This may sound too evident but in reality, it is not as such, at least in cases we have seen during our 20 years of field experience around the world! It is so unfortunate to see that in many industries around the world, it is the likelihood of corrosion, which is the main focus of managers and not the risk (we discussed the difference between these two earlier in this book).

In the next chapter, we will look at the main topic of this book, that is, microbiologically influenced corrosion (MIC) in more detail. As we have indicated several times, this book is not about the mechanisms by which MIC occurs or the details of how it is treated and monitored. This book undertakes to use the next chapter to show the readers how significant and complex MIC could be and how the way it is recognized is vastly different from all other corrosion processes: no other corrosion processes except MIC use the recognition parameters and tools for identification of planktonic and sessile bacteria. In addition to recognition, there is also another matter that within the context of our book is only unique to MIC: in addition to application of CP, use of coatings and chemical and other five forms of controlling corrosion that we discussed above in required length, there is an option of using bacteria against bacteria (biological treatment of MIC) that is the main topic of this book.

REFERENCES

In this section, we rather discussed about matters that in essence belong to the general body of knowledge of corrosion. Therefore we prefer instead of introducing individual references here, just name some resources that can have a general knowledge nature for the readers and can be addressed as "For further readings":

- For a general reading about corrosion and its electrochemistry, the following can be suggested although the interested reader can find many more:
  - R. Javaherdashti, "Microbiologically Influenced Corrosion-An Engineering Insight", Springer, second Edition, 2017.
  - P.R. Ruberg, "Handbook of Corrosion Engineering", McGraw-Hill Education, 2012.
  - Z. Ahmad, "Principles of Corrosion Engineering and Corrosion Control", Butterworth-Heinemann, 2006.
  - E. Mattson, "Basic Corrosion Technology for scientists and engineers", Ellis Horwood Publishers, 1989.
- There are numerous references on biocides, oxygen scavengers, and biocides. We took Fig. 2.8 from the reference below:
  - P. Rajeev, A.O. Surendranathan, Ch.S.N. Murthy and entitled "Corrosion mitigation of the oil well steels using organic inhibitors-a review" as published in The Journal of materials and Environmental Sciences, Vol. 3, No. 5 pp.856–869, 2012.
- About natural corrosion inhibitors, the following can be suggested:
  - Abdallah, M. (2004), "Guar gum as corrosion inhibitor for carbon steel in sulfuric acid solutions", Portugaliae Electrochimica Acta, Vol. 22, pp. 161–75.
  - Abdel-Gaber, A.M., Abd-EL-Nabey, B.A., Sidahmed, I.M., El-Zayaday, A.M. and Saadawy, M. (2006), "Inhibitive action of some plant extracts on the corrosion of steel in acidic media", Corrosion Science, Vol. 48, pp. 2765–79.

CHAPTER 3

An Introduction to Microbial Corrosion

HOW ARE MICROBIAL CORROSION, IN TERMS OF BEING AN ELECTROCHEMICAL PROCESS, AND MICROBIOLOGY RELATED?

As we mentioned earlier, for electrochemical corrosion to occur, it is necessary to have a medium through which ions can be transferred. This medium, the electrolyte, has to be conductive. The best option will then be water. It is water that, because of its conductivity, can play the role of an excellent electrolyte should it contain the necessary electrical charges that would act like an electrolyte. The meaning of this is that water per se cannot be an electrolyte and it has to have certain properties to make it the best nominee for being an electrolyte.

It follows then that if somehow we manage to make water out of its ions (the so-called demineralized water or, for short, demini water), then although we could have anode, cathode, and a metallic path that would let the electrons pass, because of having a poor or rather nonconductive water, electrochemical corrosion will not occur. This is exactly what happens in reality: the water to be used in many precise industry applications is made demini water so that the likelihood of corrosion will be expected to decrease. On the other hand, waters like seawater, for instance, are highly conductive because of their dissolved salts and therefore very good electrolytes.

It may also be a strike of luck that bacteria also prefer water for living. It is through this medium that bacteria can give/take their necessary chemicals and it is in this medium that their food is found in the best possible form they can digest. When we talk about water, we do not necessarily mean a body of water, even the moisture in the soil or in a natural gas stream within a pipeline will suffice for both MIC and electrochemical corrosion to find their suitable environment to carry on.

WHAT IS MIC AND WHAT ARE THE FOUR BASIC CHARACTERISTICS OF IT?

Microbial corrosion can be defined, and as a matter of fact "is defined," in various ways by various researchers. No matter how you define it, MIC has three features that appear as common in all of these definitions[1]:

1. MIC is an electrochemical process.
2. Microorganisms are capable of affecting the extent, severity, and course of corrosion.
3. In addition to the presence of microorganisms, nutrients and water must also be present and available to let MIC commence and continue.

   In addition to the three items above, we want to add one more item too.

4. The central concept of MIC is biofilm and its formation mechanism. However, the term biofilm is itself misleading!

We will briefly explain each of the four items in necessary length below, bearing in mind that what is being represented is just the tip of the iceberg.

MIC as an Electrochemical Process

We have repeatedly mentioned here that MIC is indeed a known corrosion process that in origin it is electrochemical. In other words, MIC is not a "type" of corrosion. In fact, the fact that MIC is not a type of corrosion has also been indicated and emphasized upon in "**Detection, Testing, and Evaluation of Microbiologically Influenced Corrosion (MIC) on External Surfaces of; Buried Pipelines; NACE Standard TM0106-2006.**" We believe that this clear and straightforward emphasis upon MIC not being a type of corrosion

Biological Treatment of Microbial Corrosion. https://doi.org/10.1016/B978-0-12-816108-1.00003-3

but a rather a corrosion process that can be explained within the known framework of corrosion control measure already used for controlling electrochemical corrosion is of great practical importance: this way it can become possible for the industry experts as well as researchers to try to find ways by which treatment of MIC could be achievable. It is also in this regard that theories such as EMIC can be better understood where bacteria due to direct electron pick-up from metal will become itself a living cathode and therefore could increase the rate of corrosion to very high values. To summarize, in order to understand MIC, one has to know two scientific majors: microbiology and electrochemical corrosion.

Diversity of Organisms Involved in Microbial Corrosion

In fact, an alternative name for MIC is biocorrosion. This, in our opinion, is more convenient that microbial corrosion, as it implies that both micro- and macroorganisms can get involved in the process of corrosion. Fig. 3.1A shows an example of when macroorganisms (that is, the organisms that can be observed with naked eye) could be involved in the deterioration process. In this figure, it can be seen that at least three corrosion mechanisms are in action: the uniform corrosion of the above water line section of the column, probably because of the marine atmosphere, the tidal zone corrosion where the zone is affected by high-oxygen and low-oxygen conditions at low tide and high tide, as well as the biofouling of the marine creatures that have been stuck on the column. We said at least three mechanisms but we do not confine it to that number. In cases as such, one thing that is of significant importance for advising a feasible corrosion management program is to realize if these corrosion mechanisms are functioning in series or in parallel. A good

FIG. 3.1 (A) An example of marine organism stuck on the tidal zone of a steel column. Their main contribution to corrosion could be via establishing differential aeration cells. (B) MIC or not MIC? (Photo: Dr. R.Javaherdashti's collection.)

mitigation program will have to revolve around this priority to be successful.

Therefore, it is not just microorganisms such as bacteria and/or archaea that can contribute to corrosion. The practical importance of this can be that, when on offshore platforms, one gets stringers clogged by marine creatures (a process known as biofouling), if one knows that it is through the formation of biofilm that serves as a "sticky tape" to larger marine life to let them grow on it, then it will make sense to first try to control biofilm formation instead of spending large sums of money on dealing with controlling biofouling.

Suitable Environment for MIC to Occur

MIC cannot take place if the following three basic requirements are not met:

- Corrosion-related bacteria (CRB)
- Nutrients
- Water/water-containing environment

Let us first start with CRB. It may sound obvious but one will be amazed to see how frequently people address a corrosion case as to be microbial (even without checking it out) just because they have seen something that might have had some resemblance with some features of MIC as they have read or overheard. A real case history as such is as follows.

As a corrosion expert, we are receiving many requests over a week in which people send the photos of the failed parts along with some process parameters that in their minds are important for us to know and then ask if the type of corrosion/failure could have been an example of MIC! An example of such is seen in Fig. 3.1B.

An inquirer sent us the pictures shown in Fig. 3.1B. He said that the figures belonged to a part of a pipe used in upstream gas and oil industry. The fluid inside that passes through the pipe contains 3% hydrogen sulfide and 5% carbon dioxide along with methane. The temperature is 120°C and the pH is 5.5. Very severe corrosion has been observed, and inside the pipe, a rather thick, black deposit has also been observed. The inquiry was whether the cause of corrosion could be MIC by sulfate-reducing bacteria! (Perhaps based on the color of the deposits!) The first question that comes to mind is that how one can talk about the possibility of MIC if no analytical method has been applied to show the existence of any bacteria, let alone sulfate-reducing bacteria? Although we do not open up the discussion that even mere existence of CRB (such as but not limited to SRB) is not a convincing reason to make us believe the case is indeed MIC.

We hope the example above has served to show our readers that in real-life industry practices, they may not even be that much concerned about the rather obvious basic requirement of existence of bacteria to confirm that the case might be MIC-related.

Another important condition for sustaining MIC is existence of nutrients in the environment. What we mean by nutrients are chemicals such as phosphorus, carbon, hydrogen, and the like.

Biofilm, A True Concept With a Wrong Name

Perhaps the most important, and in a way, the central concept of MIC is the concept of biofilm. It is essential because by forming biofilm, a great deal of corrosion will be expected to happen and one can conclude that MIC and biofilm formation are closely related.[2]

The term "biofilm" has been coined in 1978,[3] and it can be defined as "'matrix-enclosed bacterial populations' adherent to each other and/or to surfaces or interfaces."[4] To understand the concept of biofilm formation, we must understand the "states" at which bacteria can exist.

As long as bacteria can find food in the nearby bulk solution, they will prefer to swim around. In this state, they can be called as "freely swimming" or planktonic. However, if the food falls down under its weight on surfaces and stays there and bacteria "sense" that the food for them is not in the bulk but rather can be reached on surfaces, then they will drive down onto these surfaces and through a very sophisticated mechanism, details of which are still unknown to us, settle themselves on these surfaces, and become "motionless" or sessile bacteria. It is these sessile bacteria that make up the so-called biofilm and, through various mechanisms,[5] can contribute to

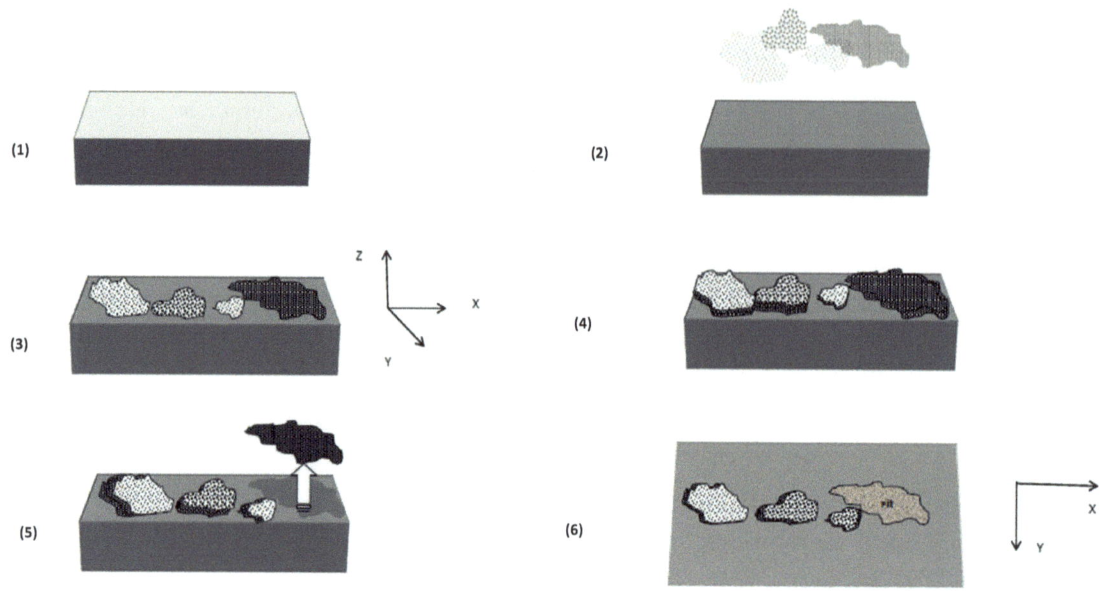

FIG. 3.2 Steps involved in biofilm buildup (construction) and deconstruction and how this process contributes to corrosion in form of pitting. (From Javaherdashti R. Corrosion and biofilm. In: Kanematsu H, Barry DM, eds. *Biofilm and Materials Science*. Springer; 2015.)

corrosion, one of these mechanisms by which corrosion can be enhanced can be through formation of spots of low and high oxygen partial pressure adjacent to each other. This can be even more enhanced by the formation of yet another form of electrochemical cells known as ion concentration cells. These formation processes of electrochemical cells would make biofilm into a rather "aggressive" entity that would increase the likelihood of corrosion. It may, however, be good to know that there is a certain time lag between the time that biofilms form and the time they start to become aggressive that in the case of filtered and unfiltered seawater may differ from weeks to months.[6] Fig. 3.2 schematically shows biofilm formation stages and how it can facilitate corrosion via, for instance, establishing differential aeration electrochemical cells. Fig. 3.3 schematically shows the way adjacent high- and low-oxygen spots can be formed.

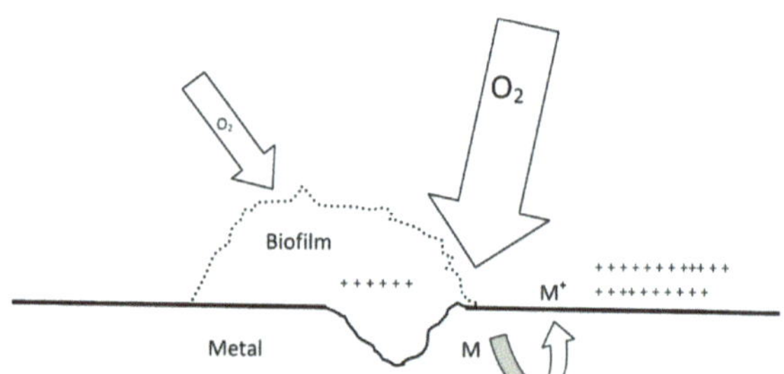

FIG. 3.3 Differential aeration cell and ion concentration cell mechanisms as operative on a biofilm-metal system. Relative size differences of the arrows represent the corresponding oxygen partial pressures. When via cathodic reactions atomic metal (M) is oxidized, it will liberate enough positively charged ions (cations, M^{+}) to shift the charge balance to yield ion concentration difference. In either case, pitting under the biofilm is inevitable. (From Javaherdashti R. Corrosion and biofilm. In: Kanematsu H, Barry DM, eds. *Biofilm and Materials Science*. Springer; 2015.)

There is also another important issue about biofilm that will make them a very confusing issue especially when it comes to industrial practices: biofilms, because of their the ever-changing structure, can easily become the focal point of all types of microorganisms, not all of which are necessarily CRB. Some of these biofilm guests could well be potentially pathogenic bacteria (e.g., Legionella) that over time could pose a serious threat to the surroundings. We will get back to this issue later in this chapter, but we want to put emphasis on the fact that by letting biofilm grow and sustain, we are actually letting not only the risk of MIC to become more serious but also the possible issues with pathogenic bacteria's sustained growth and spreading over. Biofilm control, then, is an issue to be handled by both corrosion department and HSE (health, safety, and environment).

We are not going to describe each step shown in Fig. 3.2, as we have done this before in required length before. However, it is necessary to say that, as can be seen from the figure, especially in stages 3 and 4, as long as the biofilm is managed to stay on the metallic surface, this may not expose a threat. It is true that the coverage of biofilm is not uniform and we can still get some formations of electrochemical cells; however, it is also important to see that these cells could be quite small in size and may actually not show a significant threat to the mechanical integrity of the metallic part of the asset. It is possible to think that even if small-size perforations are formed, these will be filled by corrosion products that themselves are the result of formation of other corrosion processes or the adjacent corrosion sites. This way, the anodes may be blinded by being filled with these products. However, the large-scale problem starts when the biofilm patches are removed or shifted in a significant shifting process as shown in stage 5. This can easily be leading into stage 6 where owing to formation of various electrochemical cells such as differential aeration cells (adjacent sites with different oxygen partial pressures), pitting and perforation will become a dominant process. The "holes" thus produced during this process and as a result of it will serve to enhance crack formations, and therefore, local forces will be exerted on smaller surfaces. The end result will be local increase in the stress that can well pass the yield strength of the material and thus cause its failure.

In soil, there are no freely swimming bacteria of course, in those environments the bacteria would move by mechanisms such as gliding and twitching;[7] however, the end result will still be biofilm formation by the bacteria that are successful in adhering themselves onto the metallic surface.

However, one may wonder we call the concept of biofilm right but its name wrong? We have two reasons for that:

1. Biofilm is not a film at all in the sense that it may imply a rather uniform tape covered over the material. In fact, biofilm has rather a patchy shape nature, and that is why, it can accommodate several ions or allows the passage of gases or creates spots of different oxygen partial pressure. As a matter of fact, it is these openness characteristics of biofilm that allows it to be corrosive, if it were uniform and tight, it could have been taken as a rather useful internal coating and thus slow down or even stop corrosion.
2. Not all biofilm material is organic and biological, based on the dry weight of biofilm; the nonorganic content of biofilm is around 60%–70% of its dry weight.[8]

Based on the above, we think it is now time to find another alternative name for "biofilm." What we can suggest is "Temenos," the Greek term, τέμενος, for "a sacred enclosure or precinct." The reason we are suggesting Temenos instead of biofilm is obviously because under-biofilm and outside-of-biofilm could have very different conditions, from the chemicals involved to the oxygen concentration. It is as if when "biofilm" forms, that part of the system will become functionally isolated[a] from the rest of the system. Although the bulk solution could be oxygenated, the underbiofilm patches could be rather identified with anaerobic pockets or the concentration of certain ions (such as chlorides) could be well above their concentrations in the bulk. All these

[a]The word "Temenos" is derived from the Greek "temeno" that means "to cut off."

will make biofilm to act as an enclosure that has its own characteristics.

Temenos or biofilm, what has been found about these ever-changing structures, is that they do contain various kinds of bacteria. As a matter of fact, there are many types of bacteria/archaea that can be classified as to be related to MIC. Below we will just briefly describe some of them. This must not be taken in any way that the CRB (this abbreviation is defined below) are just limited to the bacteria listed below, nor is it to be understood that we will review these bacteria per se. In other words, we will list and briefly explain these bacteria from the standpoint of the role they are playing in MIC scenarios.

INTRODUCING SOME BACTERIA THAT ARE RELATED TO CORROSION (CORROSION-RELATED BACTERIA = CRB)

The reader may wonder why we are not referring to these bacteria as "corrosive" and we rather prefer to address them as "corrosion-related bacteria," or briefly, CRB. Isn't it true that these bacteria will always cause enhancement of corrosion and the mere existence of them in a system must set an alarm to the operators? The short answer is "not necessarily"! Let us explain this a little bit more but we need to answer to two questions that are in fact "hidden" in the expectation that the so-called "corrosive bacteria" is probably a better and more correct way of addressing these bacteria than "corrosion-related bacteria":

1. Isn't it true that if you have these bacteria hanging around in your system, they are the "Cassandra" of corrosion?
2. Isn't it true that these bacteria always accelerate corrosion as measured by corrosion rates?

Let us focus first on question (1):

As we have shown in one of our research studies,[9] even if CRB are present in the system as long as we have done our corrosion management homework well and efficient, the risk of MIC will be nil. This is a finding that is also being supported by practice too; in other words, mere existence of microorganism does not necessarily mean that the case is MIC. This has been clearly stated in section 9, article 9.1. of "*NACE standard TM0212-2012-Standard Test method: Detection, Testing and Evaluation of Microbiologically Influenced Corrosion on Internal Surfaces of Pipelines*" that verbatim reads "*mere detecting viable microorganism in liquid or solid samples associated with internal corrosion does not necessarily prove that MIC has occurred.*" However, we want to look at this matter from another angle point, the pre- and proceeding corrosion process. When there is more than one corrosion process involved, it is always possible to think of these competing corrosion processes as being in series or in parallel (Fig. 3.4).

What is meant here is that sometimes corrosion processes may serve to enhance (or perhaps slow down) each other and sometimes each can go its own way without virtually having any effects on the other one. Understanding series and parallel corrosion series could have invaluable practical importance: the reason is that if corrosion processes are acting in accordance with series model, then by controlling one, it is quite likely to be able to control the other. An example is the so-called reinforced concrete corrosion. In this instance, the moisture that penetrates through concrete via its surface cracks will find its way to the core of the reinforced concrete metallic phase, which is the steel bar. Therefore, the steel is affected and rusted, and as a result of corrosion, by-products and the pressure exerted by them on the concrete (that owing to its certain crystal structure cannot withstand tensile stresses and thus cracks), more cracks are created. These additional cracks will then serve as extra corridors for moisture entrance, and loss of mechanical integrity happens and thus more enhanced rusting, and this cycle repeats itself till complete and spalling will become inevitable. In this particular example, rusting of the reinforced steel bar and concrete spalling are in series: if it is not for rusting, no extra cracks will occur and therefore no further corrosion would be expected. It is followed then by either waterproofing the concrete from outside or coating the reinforced steel bar or replacing the steel bar with a nonrusting material (e.g., composite bars), the spalling and total loss of mechanical integrity of the concrete structure will be controlled.

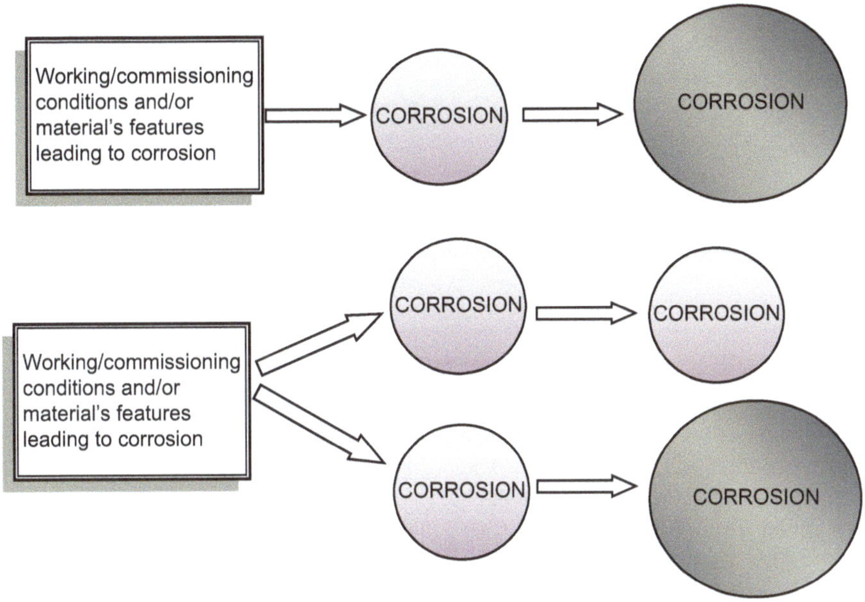

FIG. 3.4 "Series" and "parallel" corrosion processes.[10]

Based on the above, it is now easy to anticipate what parallel corrosion process could mean; in this model, two or more corrosion processes would be proceeding, which are essentially independent from each other, and thus, treatment of one does not necessarily mean that the overall corrosion problem has been resolved simply because other competing corrosion reactions will still be in action.

Now, getting back to the core question we asked earlier, we can look at MIC as a corrosion process that in many cases would appear in a parallel pattern and therefore there can be other competing corrosion reactions that could become more active and more aggressively corrosive compared with MIC alone. So, the short answer to question (1) is that mere existence of CRB may not be taken as a potential threat and therefore not all the time having these bacteria is a signal of "bad luck" of MIC.

Question (2) is even more delicate as it implies that under, practically speaking, all conditions, the bacteria that have been known to be involved in corrosion will actually accelerate it. This is not conclusive. In fact, contrary to what it may seem, even about SRB (sulfate-reducing bacteria) and IRB (iron-reducing bacteria)—that we will explain below—there is no general agreement as being "totally corrosive." For instance, experiments by Rodríguez et al.,[11] on the effect of SRB on stainless 304, lead these investigators to conclude that SRB were actually protecting metal from corrosion. In this context, Liu et al.[12] studied the possible effect of SRB on carbon steel X 80, where by increasing the biofilm coverage, MIC resistance of the metal increased. With regard to possible corrosion inhibition effects of IRB under certain conditions, researchers have done many research studies,[13,14] and in addition to personal experimenting that IRB could indeed reduce corrosion, we have addressed a good collection of it in our work.[15]

The end result of the above is that neither mere existence of bacteria can be taken as a proof of MIC nor are all the bacteria involved indeed "corrosive." Based on the above, we will use the term "CRB" to put even more emphasis on the involvement of bacteria in corrosion, which is not always a threatening issue.

There is a wide range of bacteria/archaea that could be taken as CRB and CRA (corrosion-related archaea). Table 3.1 represents schematically some bacteria/archaea that could be potentially classified as CRB/CRA.

TABLE 3.1
A Schematic Presentation of CRB/CRA Based on the Work by Al-Abbas et al.[16]

General heterotrophic bacteria/archaea	Thiosulfate-oxidizing bacteria (TOB)	Acetogenic bacteria	Slime-forming bacteria
Acid-producing bacteria/archaea	Methanogens	Nitrate-reducing bacteria/archaea	Sulfur-oxidizing bacteria/archaea
Sulfate-reducing bacteria/archaea	Nitrite-reducing, sulfur-oxidizing bacteria (NR-SOB)	Thiosulfate-reducing bacteria (TRB)	Thiosulfate-reducing archaea (TRA)

We are not claiming that in a given biofilm environment, one can find all these CRB/CRA at the same time and in the same location. However, from the types given in the table, it must be obvious to the reader that although corrosion per se is a confusing issue, MIC can make it even more complex and confusing in terms of inducing various types of life forms that can be accommodated there and the role they can play in affecting corrosion. In addition to that, we want to once more remind our readers that biofilms are not necessarily all composed of organic matter and the inorganic matter that—especially in rather nonclean industrial systems—can be accumulated; it can itself have huge detrimental effects on enhancing corrosion. Below we will briefly touch some CRB from a corrosion point of view. The reason we have selected these CRB is not because they are very important CRB but simply because they have been appearing more frequently in MIC-related descriptions of failures.

Sulfate-Reducing Bacteria

SRB are most probably the most notorious type of CRB known to industry. However, this is in fact one of the "myths" of MIC repeatedly put forwarded by self-declared MIC experts. Sulfate-reducing bacteria (SRB) that grow aerobically (no oxygen) are known for their capability of reducing sulfate ($SO^{-2}{}_4$) to sulfide (S^{-2}) where, in the absence of metallic ions such as iron, they produce H_2S gas[17] and, with ferrous ion, iron sulfide is generated. The "black rust" iron sulfide consumes the iron that was necessary to keep the original steel structure mechanically strong and sound. SRB may be found everywhere from soil[18] to seawater.[19]

These bacteria are expected to induce very high corrosion rates that in some case have been as high 10 mm per year.[20] These bacteria are so exotic to researchers, maybe because they are anaerobic (do not require oxygen for their growth) in an all-oxygen demanding environment like the earth. Since the time it was tried to explain their corrosive effects in terms of involving concepts of electrochemistry (the so-called classic theory), many rival theories have emerged to explain the unusually too high corrosion rates generated by SRB. A collection of these theories is given in Table 3.2.

As it can be seen from Table 3.2, the rather recent opinion census among experts is the last one: this recent theory—abbreviated as EMIC that stands for electrical microbiologically influenced corrosion—states that SRB can directly take up the electron that needs through nanosized tubes that—like a straw in a fizzy drink—will be used to transfer electrons from metal to the bacteria, sulfur-oxidizing bacteria. This can be schematically shown in Fig. 3.5.

With very rough terms, we can simplify explain what is seen in Fig. 3.5 as the following: SRB will be steeled onto the metal and produce back-colored iron sulfate deposits on the metal. The procedure is that SRB will, literally speaking, grow on nanotubes that will assist them in taking electrons directly to themselves via conductive metal and sulfate film. By doing so, bacteria will, practically speaking, become a cathode and the metal an anode. This will induce very high

TABLE 3.2
A Summary of Theories Put Forwarded So far for Explaining the Corrosive Effect of SRB

Classical Theory (VonWolzogen Kuhr and Van der Vlugt, 1934):	Other Theories:
• According to this explanation, the bacteria use the cathodic hydrogen through consumption by an enzyme called hydrogenase • Thus, main probable effect of SRB on corroding metal is the removal of hydrogen from the metal surface by means of hydrogenase and catalyzing the reversible activation of hydrogen	• Galvanic corrosion between the iron sulfide film and underlying steel (Stümper, 1923) • Mixed effect of hydrogenase and the FeS-Steel galvanic effect (Miller and King, 1971) • Replacing FeS with hydrogen sulfide as the cathodic reactant (Costello, mid-1970s) • The existence of a corrosive volatile phosphorous metabolite leading to observed high corrosion (Iverson, 1998) • Direct interaction of SRB and the metal to facilitate direct electron pick-up (Hang et al., 2003)

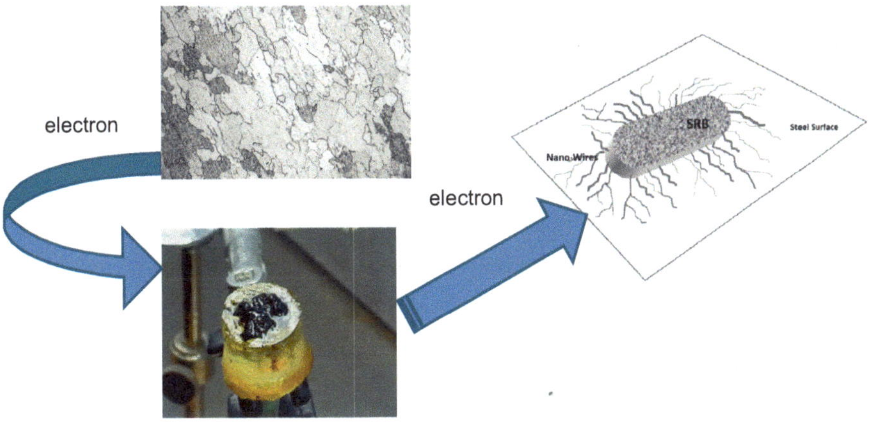

FIG. 3.5 How Can EMIC Take Place?

corrosion rates that will eventually cause accelerated corrosion of the part.

The smart reader will notice that in Table 3.2, we have not closed the list of theories that can be suggested. The main reason is that just by looking how these theories have been evolved, one can deduce that sooner or later new theories will emerge that maybe like EMIC are not just new theories but as an example of the fifth level of innovation as per TRIZ,[b] a ground-breaking paradigm in the literature of MIC.

[b]TRIZ is the abbreviation of the Russian term "теория решения изобретательскич задач" that can be translated as "the theory of inventive problem solving."

Iron-Oxidizing Bacteria

IOB utilize ferrous iron Fe^{+2} as a source of energy and are characterized by the deposition of ferric Fe^{+3} hydroxide. A common example of IOB is the *Gallionella* sp. These bacteria will oxidize ferrous to ferric and therefore this will make the initial "food" for IRB (see below). In addition, we should not forget that oxidation of iron and converting iron atom into its ionic form (ferrous) is in fact itself an anodic reaction (that is, corrosion). Therefore, these bacteria will in fact enhance corrosion by this way too. Fig. 3.6 shows an example of IOB:

Iron-oxidizing bacteria are not just a member of CRB in its negative sense: these bacteria can

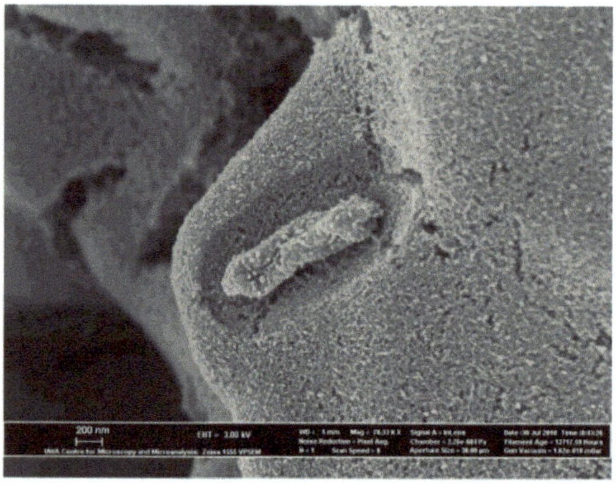

FIG. 3.6 An example of iron-oxidizing bacteria (ferrooxidans) grown on chalcopyrite.[15]

have very useful applications in mining and particularly in a method that is called "bioleaching[1]" that is applied to extract copper from its ore when routine chemical treatment methods are not feasible to be applied.

Iron-Reducing Bacteria

In simple terms, the way by which IRB can contribute to corrosion, and thus MIC, is that they convert (reduce) insoluble ferric iron into soluble ferrous ion. Although ferric ion can assist in generating protective films against corrosion, ferrous ion will be dissolved easily and fresh surface will be presented to the corrosive environment surrounding the material.[21] How can IRB be pivotal to MIC? The following could be some mechanisms by which IRB can contribute to corrosion[15]:

1. Availability of iron: Iron is not very soluble but if it is reduced to ferrous iron (which is soluble), the organic compounds can stabilize iron by chelation where, later on, that iron can "liberate" itself from the organic matter and precipitate as iron.
2. IRB are a very important part of the soil microbial community, as most of the IRB are facultative anaerobes; thus if oxygen is available, they will prefer it for their growth while maintaining also their capability of growth under anaerobic conditions too. It is estimated that in the surface layer of soil, on the average, the number of IRB could be as 10^6 cells per gram of soil. It must be reminded that as IRB are both chemoheterotrophic (organic compounds are the source of energy for them) and facultative anaerobes, their numbers within the soil's surface layer are higher than deeper levels especially if the soil is rich in organic matter at the surface level. As a result, in case their numbers in soil are reported, the depth of sampling for the organic carbon content must also be recorded.
3. Incorporation (assimilation) of iron into proteins containing, for instance, iron-sulfur: IRB are capable of making the environment suitable for SRB. In a mixed population of microorganisms in a biofilm, as oxygen is consumed, the redox (reduction-oxidation) potential starts to decrease so that nitrate, then manganic and ferric ion, and the sulfate are reduced.

Acid-Producing Bacteria—Nonorganic Producers

Sulfur-oxidizing bacteria (SOB): SOB are a family of bacteria that require oxygen for their growth (aerobic bacteria). SOB can produce sulfuric acid with very low pH values down to one even.[22] A type of these bacteria known as colorless SOB produces sulfuric acid by either of the reactions given below[23]:

$$H_2S + 2O_2 \rightarrow H_2SO_4$$

or

$$2S^0 + 3O_2 + 2H_2O \rightarrow 2H_2SO_4$$

Colorless SOB can be found everywhere in the nature and be isolated from almost every aquatic system. A genus of SOB known as *Thiobacillus errooxidans* is capable of producing 10% concentrated sulfuric acid with pH less than 2.5.[22,23]

Acid-Producing Bacteria—Organic Acid Producers

Clostridia:

Clostridia are perhaps the most dangerous of CRB about which we still do not know that much except that:

1. Like SRB, these bacteria are also anaerobic.
2. There are contradicting reports about these bacteria and their impact on corrosion.
3. They can cause very serious diseases.

The mechanism by which *Clostridia* and their various kinds can facilitate corrosion is mainly by generation of organic acids as metabolic by-products, thus lowering the pH and sustaining corrosive conditions. These bacteria are also known to produce ample amounts of hydrogen, thus increasing the likelihood of hydrogen-induced cracking (HIC). They are capable of applying at least three mechanisms that, potentially, will enhance corrosion:

1. Enhancing anodic reactions by producing acids
2. Facilitating hydrogen-induced cracking (HIC) via hydrogen generation
3. Constant availability of freshly corroding steel surface by ferric iron reduction, similar to IRB

In addition to the above, *Clostridium* sp. have been reported to be resistant to high temperatures. This thermophilic feature of *Clostridia* can make them an ideal corrosion culprit in an environment that is anaerobic and relatively hot as well as rich in nutrients necessary to sustain *Clostridia* alive and active.

These bacteria have been quoted to have caused corrosion in systems such as subsea carbon steel pipe lines, natural gas pipelines, injection systems using produced brine to displace oil from the reservoir as well as a potential problem in closed water systems that could form anaerobic environments. *Clostridia* are also very dangerous bacteria from health-related point of view.

The information mentioned above along with more detailed and updated knowledge about *Clostridia* has been given elsewhere.[15]

Possible Role of Magnetic Bacteria in Inducing Corrosion

So far, we talked about CRB about which there have been enough field and research evidence and support that we are almost certain that bacteria such as SRB, SOB, IOB, and IRB—along with those that have been mentioned in Table .3.1—do have some kind of contribution to corrosion in the form of MIC. However, the bacteria we are going to mention here are in fact bacteria about which we have not faced any official corrosion report! Nevertheless, we accept this risk and do include magnetotactic bacteria in our list of CRB. We believe–as we will explain below—magnetotactic bacteria could in fact make much contribution to corrosion even when compared with SRB should the conditions for the former become sustainable.

Among various types of microorganisms that can be found in nature, there is a class of bacteria that can be collectively referred to as "magnetotactic bacteria" or, equivalently, magnetic bacteria. These bacteria are termed magnetic because of their ability to form nanosized magnets within their "body"[24] (Fig. 3.7).[15]

The average size of each of these nanosize magnets (called a magnetosome) is about 50 nm.[25] Magnetic bacteria were discovered in the mid-1970s and they can either be aerobic or anaerobic.[26] As magnetic bacteria are capable of accumulating huge amounts of iron,[25] it has been theorized they may be an instrumental factor in inducing corrosion.[15,27,28] According to Hughes,[25] although up to 80% of magnetosome consists of magnetite (Fe_3O_4), it can highly be possible that other minerals of iron—for example, iron sulfides—may also exist in some magnetic bacteria. Javaherdashti[15] proposed for the first time the possible role of these bacteria in corrosion and thus the need to classify them as CRB as well. His proposed mechanism is based on deduction about the possible link between MIC and magnetic bacteria based on the following two fundamental questions:

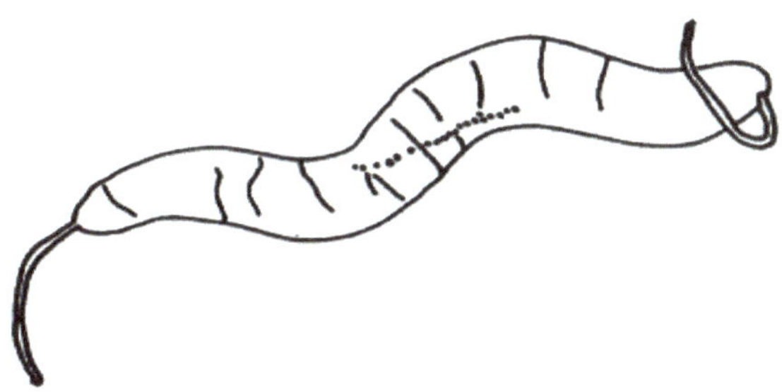

FIG. 3.7 Schematic presentation of a magnetotactic bacterium where the magnetosomes can be seen as black beads.

1. Magnetic bacteria can accumulate iron approximately 20,000- to 40,000-fold over its extracellular concentration. Between 14% and 79% by weight of the magnetosome is magnetite (Fe_3O_4). If these bacteria need this much iron, from where can they get it? One most possible way is that these bacteria must encourage the corrosion process to accumulate as much iron ions as they can. This, in turn, could simply mean magnetic bacteria would accelerate corrosion by affecting anodic reaction of metal dissolution.
2. Proposed model for magnetite biomineralization in *Magnetospirillum* species is that Fe(III) is actively taken up by the bacteria and then, it is thought to be reoxidized, resulting in magnetite production within the magnetosome. Could magnetosome formation mechanisms contribute to corrosion in the way that iron-oxidizing bacteria do by consuming ferrous ions? We mentioned above that IOB would contribute to corrosion by uptaking ferrous ions and oxidizing them to ferric ions. This by itself is an anodic reaction that would serve to increase corrosion rates. In addition to that, let us not forget that the ferric ions thus produced could always find an opportunity to be reduced by the nearby IRB.

The above are just some "scientific gut feeling" that may or may not be true about magnetic bacteria. We prefer to "bite the bullet" and keep magnetic bacteria on the list of CRB so far as hard-core evidence is found that would show they were not involved in MIC in any way.

USE OF MATHEMATICS IN MICROBIOLOGY: FUZZY MODELING FOR DISTINGUISHING CRB AND PATHOGENIC BACTERIA

Mathematics is a very powerful tool that, in case applied correctly, could create wonders. If we apply mathematics in building mathematical models, these models can then be used for, at least, two reasons: understanding the present state of a system and predicting its future state.

"Ordinary" mathematics and logic relies on binary systems. In this system, a statement is either right or wrong. If it is right, it will get a value of "1," and if it is wrong, its value will be "0." If we say, for example, "John is 35 years old AND He is old," this combines two statements of which one is right (John is 35) and the other is wrong ([John] is old). It seems that the combination of these two statements by "AND" will make a wrong result: John cannot be old but 35 years old. But what about considering 35 years old as being young? We know from our everyday experiences that definition of "young" and "not young = old" is not that easy. It is this very approach of "not being easy" and the uncertainty associated with it that makes ordinary binary logic not that useful. In fact, instead of two fixed values of 1 and 0, we can talk about a range of values that can later be expressed as percentages. Let us give another example that may be more self-expressing: we all know that most of the time, it is not that easy to say if it will rain or not. However, using new methodologies, one can easily hear on the weather forecast programs news such as "most probably tomorrow we will have rain!" What is this "most probably"? It is based on applying fuzzy logic to meteorology where by measuring some important parameters, with a certain probability, the case of weather condition for tomorrow could be predicted. Fuzzy logic, proposed by Iranian-American scientist Lotfi Askarzadeh (Zade) in the 1960s, is now the required theoretical basis for essentially any device that is branded as "smart." Fuzzy logic, in simple terms, is the mathematics of uncertainty.[3]

Although CRB can alter the severity of corrosion in various ways, confusing as it is, some CRB can be pathogenic too. For example, the pathogenic role of *Clostridia in Gangaria* has also been reported.[29] In addition to *Clostridia*, *Pseudomonas* are also a group of CRB that although they can enhance corrosion by rapidly forming biofilm under which various corrosive reactions will be facilitated, they can also be a health issue in several cases.[c,30].

---

[c]There are a lot of publications about what fuzzy logic and fuzzy calculus are all about. We have introduced many of such publications in our previous works that we have also referenced here.

It is also possible to think that there can be other bacteria that in their natural path of getting energy are capable to induce harm to both human body (pathogenic) or human-made structures (CRB). If this is true, this will mean that for both microbial corrosion and medical microbiology, studying of dual-functioning CRB (corrosion and disease) is a greenfield for doing research. This classification of bacteria into PB (pathogenic bacteria) and CRB (corrosion-related bacteria) does not necessarily mean that *all* bacteria are either PB or CRB. This is certainly oversimplification and certainly not true. However, classifying bacteria into PB and CRB simply means that some CRB could also be PB and some PB could also play a role that would make them suitable to be considered as CRB. In other words, our classification does not apply to all bacteria species but to only those that can be both CRB and PB at the same time. It will also serve us to mathematically describe the expected behavior from these bacteria in a more convenient way.

In the past in a series of works, we have applied fuzzy logics and reasoning to successfully predict the risk of MIC.[31–33] Here, we will develop the same methodology with augmenting some fundamental elements that would enable us to define mathematically what "more or less corrosive bacteria" or "more or less pathogenic bacteria" would mean. This first-ever modeling methodology will enable researchers in a multidisciplinary arena to look for new solutions and methodologies.

Eight Fundamental Assumptions

The model we would like to propose for the assessment of the health risk associated with corrosion-related bacteria will be based on fuzzy calculations.

The fuzzy algorithm that we will use in our approach here is based on the following initial conditions:

1. Corrosion-related bacteria (CRB) do contribute to corrosion via biofilm formation.[5]
2. Biofilm formation is a process that rests on a variety of factors that can be grouped into physicomechanical factors and chemical factors.[34]
3. In the context of this modeling, pathogenic bacteria (PB) will mean those bacteria, which are contributing to human health problems via infection mechanisms.
4. PB will do the infection process by "adherence" onto the healthy cells and then "invasion."
5. Although invasion includes essentially chemical factors, many of these factors cannot be modeled and mathematically quantified. However, adherence does include various measurable physical as well as chemical parameters.
6. CRB, which are PB as well, are collectively referred to as CRB-PB.
7. The fuzzy algorithm describing CRB-PB is a combination of fuzzy mode/description of CRB corrosive action via biofilm formation and fuzzy model/description of adherence for PB via infection mechanisms.

General Methodology

We will define a general set of features related to both biofilm formation (main assumed MIC mechanism by CRB) and infection (main assumed infection mechanism by PB). Then we will define some subsets related to each of these general sets. A fuzzy membership function for each set is defined to arrive at composite function of membership functions. By defining the composite functions, fuzzy rules to characterize the environment and its important parameters are defined. A fuzzy method known as "generalization of compositional rule of inference" is utilized in this study. In this method, a fuzzy rule is transformed into a general form of multiantecedents (inputs) and consequents (outputs). Also "Kosko decomposition method" for decomposing a fuzzy rule and Mamdani minimum fuzzy implication will be used. Utilizing "Mamdani minimum fuzzy implication," the minimum value of membership functions of the given fuzzy sets is calculated. Then, by "maximum-minimum technique," first the minimum values of membership functions are calculated. After that, among the selected minimum values, the maximum value is picked up. If, in any case, the membership functions of some elements are equal, one of the functions is chosen. More details related to evaluation of fuzzy methods and their applications are found in other references. More detailed explanation of these methods has been given elsewhere.[31]

The above methodology will be applied to both CRB and PB categorization sets. After obtaining the general fuzzy membership functions for CRB and PB, using appropriate fuzzy rules a composite fuzzy membership function for CRB-PB will be defined.

CRB FUZZY MODELING

Three sets are defined as follows:

$G = \{G_j\}$, $j = 1, 2, 3, \ldots N$

$C = \{C_j\}$, $i = 1, 2, 3, \ldots P$

$A = \{A(i)\}$, $i = 1, 2, 3, \ldots P$

Set "G" contains all detectable, measureable, and assessable features of any bacteria. Each "G" parameter will show a parameter of bacteria that helps classifying them. These parameters will be, but not limited to, reaction to Gram stain, morphology, metabolic by-products, temperature sensitivity (mesophilic, thermophilic, ...), oxygen requirement, if anaerobic whether the bacteria are using anaerobic respiration or fermentation and the like. Each of these parameters itself can be the product of a given mathematical model, for example. It is possible, for instance, to define a weight-based system by which having certain morphology (e.g., bacillus) has a weight (to which a number is assigned) that will make it different from having coccus morphology.

Set "C" measures required ranges of set G that will render the bacteria to become CRB. For example, if the bacteria of interest are *Pseudomonas*, then their capability in rapid forming of biofilm will make them a member of CRB. Alternatively, if a bacterium such as *Clostridia* is capable of producing ample amounts of hydrogen, then it is highly likely that hydrogen-induced cracking (a serious mode of failure) will occur in the material. The two examples above may serve to show that a natural feature already acquired by bacteria will serve to induce corrosion and thus render the material susceptible to corrosion. Also, it must be noted that the biofilm quantity and the amount of hydrogen are parameters that, in this particular example, must be above a certain range to render it a "dangerous" factor from a corrosion point of view.

Set "A" measures fuzzy probability of each member of the set G to become a member of the set C. In other words, A would measure would measure fuzzy possibility of risk of corrosion (MIC) within the given parameters.

Defining Membership Function for CRB

In biofilm formation, two sets of factors are of importance: physicomechanical factors and chemical factors. In our methodology, we will consider each set of factors separately but linked with each other.

Physicomechanical factors

Assume that there are various U features that can favor biofilm formation and not necessarily be related to each other (such as the surface temperature of the metal and its alloying elements). We may assume that for each G_j, there is a physicomechanical feature such as K so that $K = 1, 2, 3, \ldots U$.

When G_j is considered for a special physicomechanical feature of the environment such as K, it may also be assumed that the parameter will be a random variable such as x(j,K) that obeys a normal distribution function. For any factor that can help biofilm formation and is expressed as C_i, one may assume that m(i,K) and M(i,K) are, respectively, the permissible minimum and maximum thresholds for the physicomechanical feature K to be expressed by C_i. This will be translated as Eq. (3.1) in terms of Fuzzy probability function:

$$F_{A(i,K)}(G_j) = \mathrm{Prob}(m(i,K) \cdot x(j,K) \cdot M(i,K)) \qquad (3.1)$$

where $K = 1, 2, 3, \ldots, U$; $i = 1, 2, 3, \ldots, M$; $j = 1, 2, 3, \ldots, N$.

Eq. (3.1), in terms of a membership function $F_{A(i,K)}$ (G_j), defines the fuzzy likelihood of an existing physicomechanical feature such as K from the range of physicomechanical features G_j to become an element of C_i, suitable for biofilm formation. For instance, "surface roughness" is a physicomechanical feature of that is indeed a very important factor for bacteria adhesion onto the surface[35]

Chemical features of the environment

Assume that there are various V chemical features that can favor biofilm formation and not necessarily be related to each other (such as the TDS [total dissolved solids] of the bulk water and presence of some alloying elements). We may assume that for each G_j, there is a chemical feature such as L so that L = 1, 2, 3, ...,V.

When G_j is considered for a chemical feature of the environment such as L, it may also be assumed that the parameter will be a random variable such as x(j,L) that obeys a normal distribution function. For any factor that can help biofilm formation and is expressed as C_i, one may assume that m(i,L) and M(i,L) are, respectively, the permissible minimum and maximum thresholds for the chemical feature L to be expressed by C_i.

Therefore, the related membership function, in terms of fuzzy probability function, can be shown as Eq. (3.2):

$$F_{A(i,L)}(G_j) = \text{Prob}(m(i,L)\cdot x(j,L)\cdot M(i,L)) \quad (3.2)$$

where L = 1, 2, 3, ...; V, i = 1, 2, 3, ...; M, j = 1, 2, 3, ..., N.

Eq. (3.2), in terms of a membership function $F_{A(i,L)}$ (G_j), defines the fuzzy likelihood of an existing chemical feature such as L from the range of chemical features G_j to become an element of C_i, suitable for biofilm formation.

Fuzzy composite functions for CRB

Eqs. (3.1) and (3.2) define how close the value of a given physicomechanical and/or chemical feature can be to the range of risky values for becoming eligible for biofilm formation. Now these membership functions need to be defined as a single function in accordance with fuzzy functions. In other words, a composite function $F_{A(i)}$ must be defined as a function of both $F_{A(i,K)}$ (Gj) and $F_{A(i,L)}$ (G_j).

A composite function for each GjεG can be defined for physicomechanical features and chemical features of the environment, respectively, as Eqs (3.3) and (3.4):

$$F_{K(i)}(G_j) = \underset{K}{Max}\left\{F_{A(i,K)}G_j\right\} \quad (3.3)$$

$$F_{L(i)}(G_j) = \underset{L}{Max}\left\{F_{A(i,L)}G_j\right\} \quad (3.4)$$

Eqs (3.3) and (3.4) explain that among the membership functions for each set of the environmental features, the maximum values must be picked up. The fuzzy subset $A_{K(i)}$ (a member of G) defined by the membership function $F_{K(i)}$ (Gj) shows that with what (fuzzy) probability a certain range of physicomechanical features of the environment can have the conditions that will render them dangerous enough to promote biofilm formation, as indicated by C_i. Likewise, the fuzzy subset $A_{L(i)}$ (a member of G) defined by the membership function $F_{L(i)}$ (Gj) shows that with what (Fuzzy) probability a certain range of chemical features of the environment can have the conditions of becoming a promoter for biofilm formation, as, again, indicated by C_i. It must be noted that the values of both $F_{A(i,K)}G_j$ and $F_{A(i,L)}G_j$ can be assumed not to be arbitrary variables, which are independent of each other.

Eq (3.5) defines the fuzzy membership function, $F_{A(i)}$, for the fuzzy subset A(i) (belonging to A) in such a way that it can measure the fuzzy possibility, S_i (belonging to S) of a combined range of physicomechanical and chemical features G_j (belonging to G) for becoming vulnerable to the value necessary for biofilm formation.

Assuming $G_j\varepsilon G$:

$$F_{A(i)}G_j = \begin{cases} \text{If Max}\left\{F_{K(i)}Gj, F_{L(i)}Gj\right\} = 0 \; then 0 \\ \text{If Max}\left\{F_{K(i)}Gj, F_{L(i)}Gj\right\} \neq 0 \; then \beta_1 F_{T(i)}Gj + \beta_2 F_{W(i)}Gj \end{cases} \quad (3.5)$$

where $\beta_1 + \beta_2 = 1$.

Eq (3.5) will cover the way by which CRB corrosive action may be described in terms of involving physicomechanical as well as chemical factors.

PB FUZZY MODELING

Three sets are defined as follows:

$G = \{G_j\}$, j = 1, 2, 3,...N
$S = \{S_i\}$, i = 1, 2, 3,...M
$B = \{B(i)\}$, i = 1, 2, 3,...M

Set "G" is as mentioned previously. Set "S" measures required ranges of set G that will render the bacteria to become pathogenic bacteria. Infection by itself can be thought of being composed of two series steps: "attachment" or "adherence" of the PB onto the host cell and "invasion" that can be thought of the process through which the PB enter into the host cells and propagate through healthy cells and thus make them sick. We will only concentrate on the "adherence" phase here because without adherence, it will be quite unlikely for the invasion phase to happen and thus pathogenesis will be promoted. There are various factors that will affect adherence such as host cell surface hydrophobicity, host cell net electrical charge, interactions between the binding molecules on bacteria (ligands), and host cell receptors via means physically installed on the bacteria such as "pili" or "fimbriae."[36] The first two factors may be explained by electrostatic forces, whereas the last factor (interaction of the PB bacteria with the cell) via physical structures—more or less—can be classified as mechanical entanglement of these structures. Therefore, we can define two factors: physicoelectrostatic factors and mechanical factors.

Set "B" measures fuzzy probability of each member of the set G to become a member of the set S. In other words, B would measure fuzzy possibility of risk of bacterial pathogenesis within the given parameters.

Physicoelectrostatic Factors

Assume that there are various P features that can favor PB adherence and not necessarily be related to each other (such as net electrical charge on the host cell and entropic nature of cell surface hydrophobicity) (Silverstein, 1998). We may assume that for each G_j there is electrostatic feature such as T so that $T = 1, 2, 3, \ldots P$.

When G_j is considered for a special physicoelectrostatic feature of the environment such as T, it may also be assumed that the parameter will be a random variable such as x(j,T) that obeys a normal distribution function. For any factor that can help PB adhesion and is expressed as S_i, one may assume that r(i,T) and R(i,T) are, respectively, the permissible minimum and maximum thresholds for the physicoelectrostatic feature T to be expressed by S_i.

The above will be translated as Eq. (3.6) in terms of fuzzy probability function:

$$F_{B(i,T)}(G_j) = \mathrm{Prob}(r(i,T).x(j,T).R(i,T)) \quad (3.6)$$

where $T = 1, 2, 3, \ldots, P$; $i = 1, 2, 3, \ldots, M$; $j = 1, 2, 3, \ldots, N$.

Eq. (3.6), in terms of a membership function $F_{B(i,T)}$ (G_j), defines the fuzzy likelihood of an existing physicoelectrostatic feature such as T from the range of physicoelectrostatic features G_j to become an element of S_i, suitable for PB adherence to the host cell surface.

Mechanical Factors

Assume that there are various Q mechanical features that can favor PB adherence and not necessarily be related to each other. We may assume that for each G_j there is a mechanical feature such as W so that $W = 1, 2, 3, \ldots, Q$.

When G_j is considered for a mechanical feature of PB such as W, it may also be assumed that the parameter will be a random variable such as x(j,W) that obeys a normal distribution function. For any factor that can help adherence and is expressed as S_i, one may assume that r(i,W) and R(i,W) are, respectively, the permissible minimum and maximum thresholds for the mechanical feature W to be expressed by S_i. For example, the mechanical strength of the bacterial pili structures could be an example because as a material, these pili would have certain mechanical properties that would change from bacteria to bacteria or it may even change from pili to pili over the surface of a given bacterium.

Therefore, the related membership function for the above, in terms of fuzzy probability function, can be shown as Eq. (3.7):

$$F_{B(i,W)}(G_j) = \mathrm{Prob}(r(i,W).x(j,W).R(i,W)) \quad (3.7)$$

where $W = 1, 2, 3, \ldots, Q$; $i = 1, 2, 3, \ldots, M$; $j = 1, 2, 3, \ldots, N$.

Eq. (3.7), in terms of a membership function $F_{B(i,W)}$ (G_j), defines the fuzzy likelihood of an existing mechanical feature such as W from the range of mechanical features G_j to become an element of S_i, suitable for PB adherence.

Fuzzy Composite Functions for PB

Eqs. (3.1) and (3.2) define how close the value of a given physicoelectrostatic and/or chemical feature can be to the range of risky values for becoming eligible for PB adherence. Now these membership functions need to be defined as a single function in accordance with fuzzy functions. In other words, a composite function $F_{B(i)}$ must be defined as a function of both $F_{B(i,T)}$ (G_j)and $F_{B(i,W)}$ (G_j).

A composite function for each $G_j \varepsilon G$ can be defined for physicoelectrostatic features and chemical features, respectively, as Eqs. (3.8) and (3.9):

$$F_{T(i)}(G_j) = \underset{T}{Max}\left\{F_{B(i,T)}G_j\right\} \tag{3.8}$$

$$F_{W(i)}(G_j) = \underset{W}{Max}\left\{F_{B(i,W)}G_j\right\} \tag{3.9}$$

Eqs. (3.8) and (3.9) explain that among the membership functions for each set of the environmental features, the maximum values must be picked up. The fuzzy subset B $_{T(i)}$ (a member of G) defined by the membership function $F_{T(i)}$ (G_j) shows that with what (fuzzy) probability a certain range of physicoelectrostatic features can have the conditions that will render them dangerous enough to promote PB adherence, as indicated by S_i. Likewise, the fuzzy subset $B_{T(i)}$ (a member of G) defined by the membership function $F_{L(i)}$ (G_j) shows that with what (fuzzy) probability a certain range of chemical features of the environment can have the conditions of becoming a promoter for adherence, as, again, indicated by S_i. It must be noted that the values of both F $_{B(i,T)}G_j$ and $F_{B(i,W)}G_j$ can be assumed to be not arbitrary variables, which are independent of each other.

Eq. (3.10) defines the fuzzy membership function, $F_{B(i)}$, for the fuzzy subset B(i) (belonging to B) in such a way that it can measure the fuzzy possibility, Si (belonging to S) of a combined range of physicoelectrostatic and chemical features Gj (belonging to G) for becoming vulnerable to the value necessary for PB adherence onto the host cell.

$$F_{B(i)}G_j = \begin{cases} \text{If Max}\left\{F_{T(i)}\text{Gj}, F_{W(i)}\text{Gj}\right\} = 0 \textit{ then } 0 \\ \text{If Max}\left\{F_{T(i)}\text{Gj}, F_{W(i)}\text{Gj}\right\} \neq 0 \textit{ then } \alpha_1 F_{T(i)}\text{Gj} + \alpha_2 F_{W(i)}\text{Gj} \end{cases} \tag{3.10}$$

where $\alpha_1 + \alpha_2 = 1$.

Eq. (3.10) will cover the way by which PB adherence onto the host healthy cell may be described in terms of involving physicoelectrostatic well as chemical factors.

DISCUSSION

In previous sections, we could define the membership functions for corrosion-related bacteria (CRB), designated as $F_{CRB} = F_{A(i)}G_j$, and pathogenic bacteria (PB), designated as $F_{PB} = F_{B(i)}G_j$. The aim is to define a membership function that will relate any given set of bacteria-specific features (the set G) to the set of bacteria that could have both corrosion and pathogenic features (such as *Clostridia* or *Pseudomonas*). The membership function for bacteria that would show both pathogenic and corrosion-related features can be designated as $F_{CRB\text{-}PB}$. Fig. 3.8 schematically presents the relationship between CRB, PB, and CRB-PB membership functions.

To best define $F_{CRB\text{-}PB}$, we can use approximate fuzzy reasoning, by expending the related sets for physicomechanical factors, chemical factors, physicoelectrostatic factors, and mechanical factors.

We start by our reference set of measureable properties G as below:

$$G = \{G_j\},\ j = 1, 2, 3, \ldots N.$$

For CRB, we defined physicomechanical (K) and chemical (L) sets as defined below:

$$K = \{1, 2, 3, \ldots U\}$$
$$L = \{1, 2, 3, \ldots, V\}$$

For PB, we defined physicoelectrostatic (T) and mechanical (W) sets as defined below:

$$T = \{1.2.3...., P\}$$
$$W = \{1.2.3...., Q\}$$

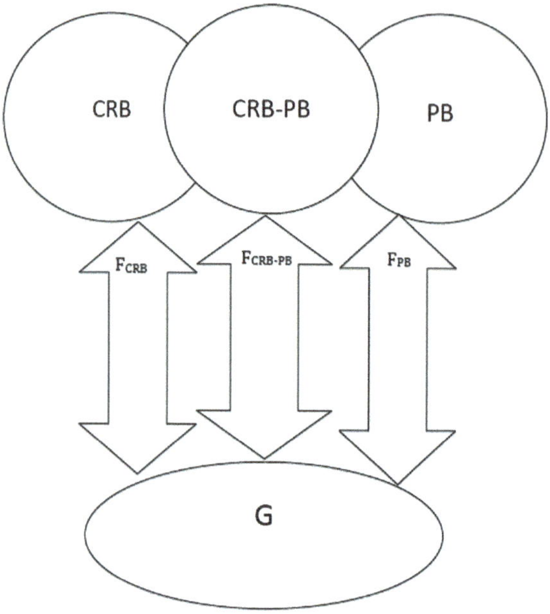

FIG. 3.8 The set G that contains all measureable features of bacteria. If these features can have values that will render them as corrosive, then the membership function F_{CRB} will define how the bacteria can be classified as corrosion-related. If, on the other hand, the bacterial features as referenced to in G are pathogenic, then the membership function F_{PB} will define the membership function to classify the bacteria as pathogenic. F_{CRB-PB} will serve as the membership function to define the fuzzy probability by which bacteria can have both corrosion-related and pathogenic features.

As such, for the reference set G, we will define the expanded fuzzy set of membership's functions as below, Eq. (3.11):

$$F_{CRB\text{-}PB} = \left\{F_{A(i)}(G_1), F_{B(i)}(G_1), F_{A(i)}(G_2), F_{B(i)}(G_2), F_{A(i)}(G_3), F_{B(i)}(G_3), \ldots, F_{A(i)}(G_j), F_{B(i)}(G_1)\right\} \tag{3.11}$$

Therefore, as per rules of fuzzy reasoning for defining "more or less," we define "bacteria which are more or less X" as below where the underlined term shows the fuzzy nature of reasoning, Eqs. (3.12) and (3.13):

"Bacteria which are more or less corrosive"

$$= \left\{F^2_{A(i)}(G_1), F_{B(i)}(G_1), F^2_{A(i)}(G_2), F_{B(i)}(G_2), \ldots F^2_{A(i)}(G_j), F_{B(i)}(G_1)\right\} \tag{3.12}$$

"Bacteria which are more or less pathogenic"

$$= \left\{F_{A(i)}(G_1), F^2_{B(i)}(G_1), F_{A(i)}(G_2), F^2_{B(i)}(G_2), \ldots F_{A(i)}(G_j), F^2_{B(i)}(G_1)\right\} \tag{3.13}$$

By applying the above methodology, with a fuzzy probability, it is possible to estimate the features of the bacteria in terms of being more or less corrosive and/or pathogenic.

SUMMARY

Some of the bacteria that induce microbiologically influenced corrosion (MIC) could also be pathogenic. Some examples of these bacterial types are *Clostridia* and *Pseudomonas*. Based on our previous successful Fuzzy models and their application into MIC cases, we have developed a model that as an algorithm can differentiate between pathogenic and corrosion-related bacteria and, by applying fuzzy reasoning, define "more or less" pathogenic or "corrosion-related" bacteria terminologies. This is a first-ever approach toward application of fuzzy calculations and reasoning to create a mathematical description of such bacteria.

HOW IS MIC RECOGNIZED? A SHORT REVIEW OF CULTURE-DEPENDENT AND CULTURE-INDEPENDENT METHODS AND THEIR PROS AND CONS

Culture-Dependent Methods

MIC is different from other corrosion processes in that not only one has to have means to cure and monitor it but also that it must first be recognized and indeed recognized correctly. It is not a rare practice to come across cases where the corrosion was not due to the activity of CRB; nevertheless, it was wrongly attributed to them. The opposite scenario is also available: cases where MIC was the main failure mechanism however due to incorrect recognition of the root cause of failure (for which existence of CRB is a necessary, but not enough, condition), the investigation had not been carried out thoroughly and in accordance with required principles. Therefore, it is of quite high importance to not only be convinced to include recognition in the essential trio of recognition-treatment-monitoring but also do the recognition task correctly. What we mean by "correctly" here is that one has to realize the pros and cons of the available methods and also that these methods are themselves tools that like all tools can only be useful if used correctly.

The issues related to MIC management must start with recognition phase, to be followed by various treatment strategies and tactics to be later followed by constant monitoring so that one can make sure that the previous two steps have been taken and applied correctly. In this section, we are not covering the techniques and technologies that are available for treatment and monitoring of MIC. This, with no doubt, must not be taken as treatment and monitoring being not important or, even worse, are simple matters: while still in some industries operators candidly think that by simply adding bleach, they can solve the problem of MIC and therefore close the file of MIC treatment, the whole MIC treatment technologies and the reasons behind them are a very sophisticated world.

The issues related to MIC are to be done in the order that we have schematically shown in Fig. 3.9. In other words, without a correct recognition, it will be a very risky business to prescribe treatment and monitoring strategists. Only when treatment was selected and applied, the required monitoring technologies must be selected or designed to make sure that the treatment strategy we have picked up has been the best possible option.

In Fig. 3.9, one of the most important initial steps in dealing with MIC is its recognition. In other words, it is quite significant to find out if any CRB/CRA is around. As we have stated earlier, mere existence of CRB/CRA cannot be taken as the final verdict about the corrosion case really being MIC in nature. However, if we can show that CRB/CRA were indeed found and that they were active

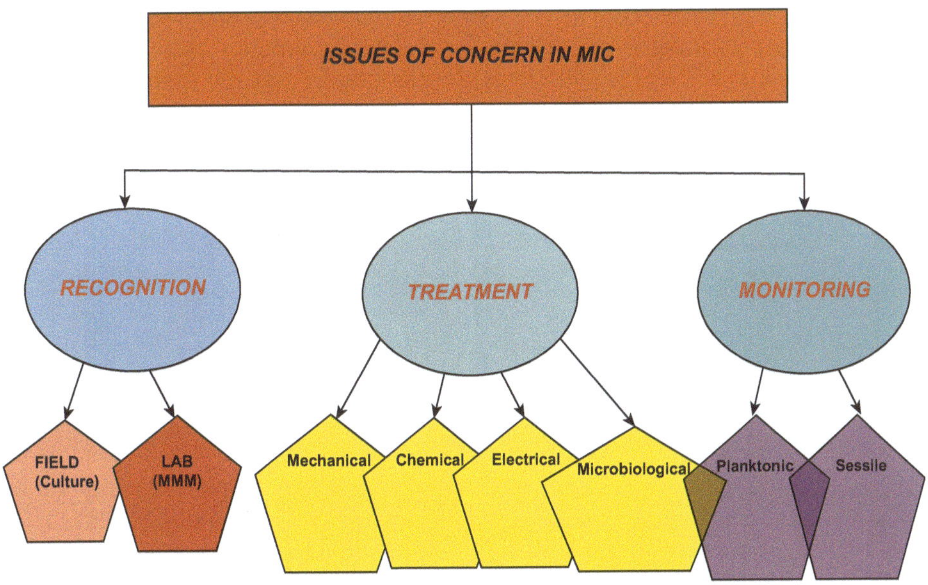

FIG. 3.9 Three important action categories to be taken seriously when dealing with any MIC case 15.

and this was coupled with other chemical and microbial evidence about the case indeed being MIC-related, then recognition of CRB/CRA can in fact be a very important step toward defining the corrosion case being indeed MIC-linked.

Recognition methods in a very wide categorization are divided into two groups: field methods and the so-called laboratory (lab) methods. This division is not a clear-cut categorization as so-called field tests are in principle all culture-dependent methods, which essentially are lab-based themselves too; nevertheless, we consider them as field tests and do not count them as lab methods.

The main reason for this confusion comes in fact from not properly understanding the way by which distinction between the two is made. We call field tests as such simply because they are not carried out in the "familiar" space of a laboratory. The field tests are carried out by people whose training in microbiology is at minimum level and categorically are performed on the field. The field tests, as a rule, use tubes that have been filed with a nutrient suitable for the growth of a certain CRB. This medium can be either liquid, gel-like, or in newer versions powered form. Based on the features of these tubes, they may be filled with either susceptible water or solid samples (Fig. 3.10).

No matter what state of matter the growth medium is in, the principle will be the same as a culture medium: certain chemicals that have been tried and known to be used by certain bacteria are added in accordance with a certain recipe.

Normally, these chemicals do include a source of energy (carbon source) and if the kit has been designed for anaerobic (oxygen-free) conditions, some oxygen scavengers may be also added or already in the medium. (In Fig. 3.10 (left), after collecting the samples for anaerobic SRB, one had to put a layer of sterile liquid paraffin along with a certain small pill that would be dissolved to produce a cushion of gas both serving to limit the ingress of oxygen.) The analogous of such field tests (or rather, rapid tests) are the familiar Petri dishes used in microbiology laboratories for preparing cultures for growing certain bacteria, including CRB.

Because in all field tests, one uses certain "food" (=culture) to grow certain bacteria, they are collectively referred to as "culture-dependent methods." As we have emphasized so far several times in these paragraphs, culture-dependent methods can only detect CRB and not CRA. In addition to this, no cultures can detect all CRB, which means that for each CRB there is a specific culture that must be prepared and used. However, culture-dependent methods are rather simple and compared with culture-independent methods, less sophisticated and cost-effective.

One should not also forget that these rapid tests usually test the planktonic population and

FIG. 3.10 (Left) A commercial kit for growing CRB within a gel-like medium. (Right) Another commercial kit for growing CRB within a liquid medium. (Photo: Dr. R. Javaherdashti Collection.)

not necessarily the sessile population of CRB. Therefore, one should always have in mind the two relatively significant short comings of culture-dependent methods, either used as field tests or actually carried out in a laboratory; these methods *cannot*

- detect CRA (corrosion-related archaea)
- detect sessile CRB (corrosion-related bacteria) population (therefore, the outcome of CRB population is expressed as cell/ml and not cell/cm^2)

In all of these field tests, the relative risk of aggressiveness of the measured CRB is expressed as a function of the number of the bacteria, if, for example, you reach a certain population within 2 days rather than 6 days, you will get a higher index of aggressiveness. This rather linear relationship between planktonic bacterial numbers and corrosion risk is not always true, especially for SRB 15. Also, a good contribution comes from sessile bacteria (biofilm) rather than planktonic bacteria, and therefore, just measuring populations of planktonic bacteria is not useful in setting the limits to evaluate the severity of MIC in a system.

As we see from the above, culture-dependent methods do suffer from significant drawbacks that can hardly compensate their advantages such as ease of doing and cost-effectiveness. However, we are not saying here that these methods are of no use at all; in fact, the point we are trying to make here is that one should not put all his/her eggs in the basket of culture-dependent methods only.

Culture-dependent methods, and particularly field tests, are essential parts of any corrosion management program whose aim is to recognize if the corrosion case is indeed related to MIC. Field tests, for that matter, have revolutionized the way field technologists were looking at MIC. It is thoroughly understood that we have such and such CRB in our system that most of the time it can serve as a wake-up call to the operators about the imminent danger.

There is also one more aspect of culture-dependent methods that must be taken into serious consideration: If one has to do it anyway, he has two choices: either do the on-field test kits or collect enough samples to be sent it later to a test house for further culture-based investigation and studies. If the latter is the inevitable, then recent findings show that it is much safer to do the inoculation right on the spot and then send it to the test house.[37] The importance of this is that in case the former option is selected, that is to say, use of commercial rapid test kits is the choice, all these kits in their accompanying vials have the necessary medium to keep the bacteria alive and even if the kits are designed for a specific type of CRB, the growth medium in it will function much more than a general purpose sustaining medium but a specific growth medium that has been formulated for that particular type of CRB.

The most popular type of these specific field tests is available for SRB, in addition to some other CRB that obviously do not include all of them. If the kit comes in specific type for a specific CRB, then there are some signatures (odor, appearance, color, etc.) that can be taken as the clues to ensure if that specific type of CRB exists in the system or not. For specific case of SRB, these signatures normally include a rotten egg (hydrogen sulfide) smell and blackening of the medium (as a result of reacting the liberated hydrogen sulfide with the iron that exists in the medium as an additive to form the black-colored iron sulfide).

Culture-Independent Methods

Culture-independent methods, or alternatively molecular microbiology methods (MMMs), are more sophisticated—in comparison with culture-dependent methods— and much more accurate technologies that are currently available.

In layman's terms, MMMs detect certain genes that can be assigned to certain CRB. This means that if SRB has a certain gene that is specifically a fingerprint of this CRB, then by establishing a one-to-one relationship between the number of this gene and the number of bacteria that have that particular gene (that is, SRB), we can enumerate the SRB in our samples.

TABLE 3.3
Selected MMMs (Culture-Independent Methods) That Are Also Applicable in MIC Studies

Polymerase chain reaction (PCR)
Denaturing gradient gel electrophoresis (DGGE)
Fluorescence in-situ hybridization (FISH)
Microbial diagnostic microarray (MDM) (gene chips)
Metabolic activity measurement techniques such as ATP measurement
Next-generation sequencing (NGS)

MMMs are more accurate in comparison with traditional culture-dependent methods. The main thing here for corrosion technologists is not the sophisticated technologies involved, but the reliable precision that can be expected from MMMs based on current knowledge and its boundaries:

Culture-independent methods (MMMs), a selection of which is enlisted in Table 3.3, have the following features that can distinguish them from their culture-dependent counterparts[15]:

- MMMs highlight the dominant bacteria in a natural sample (an ecological system) that contains other species as well, with no worries regarding the serious limits of common viable counting methods.
- MMMs determine the relative proportion of CRB within the whole bacterial community.
- MMMs evaluate the population size changes imposed by using biocide or nutrient modifications.

Except FISH, methods such as qPCR and DGGE do count dead cells too. This is very important from the standpoint of MIC treatment strategies: one must be careful not to take the numbers he/she reads as the real number of CRB and based on prescribed biocide dosage, for example. It is prudent to use other techniques that can also help us in determining the actual size of the population; in this regard, culture-dependent methods that only measure living cells or applying microscopic methods such as confocal laser scanning microscope (CLSM) can distinguish between dead and living cells[38] in a given sample.

"ZUGZWANG EFFECT" IN MIC

Google defines "Zugzwang" as *"a situation in which the obligation to make a move in one's turn is a serious, often decisive, disadvantage"*. "Zugzwang" is a term in Chess. It can be interpreted simply as "any move you do will be useless" or "you are doomed to accept that you can't do anything." Zugzwang as a concept can also be applied to MIC in the context that is more often seen in industries.

To put it in the right context, it is necessary to start with a real example: this author once was called to a power plant to talk about corrosion in fire water rings and how it can prevent.[d] This power plant had three fire water rings, two of which were completely clogged with corrosion products and were of no use any more. The third ring had severe leakages so that it was losing about 3000 L of water per day. In this situation, the first two rings that are literally out of service can be taken as an example of "Zugzwang effect" in the sense that it is not economical to try to open up the whole ring because it will obviously involve costs such as excavation (most rings are buried), replacing the rings, the man-hours to be spent for these activities, and the like expenses. On the other hand, nor is it feasible to think of any MIC treatment anymore because it is quite clear that it is now too late to think of any possible

[d]"Control" of corrosion is different from its "prevention": When we use "corrosion control," we mean that corrosion has already started and we have diagnosed it and our aim here is not to let it spread nor cause leakage. On the other hand, when we say "corrosion prevention," it means that we have some plans in place that will lower the likelihood of corrosion being started to very low levels. The "corrosion control" strategy and tactics are surely different from "corrosion prevention." For example, whereas in a corrosion prevention scheme we may have the freedom of material selection and using materials that are more corrosion resistant, in a corrosion control scheme, we may not have this option and thus we would have to find alternatives that are both feasible and applicable.

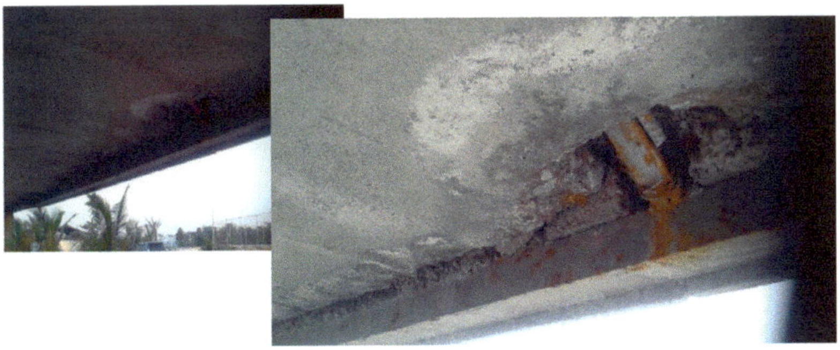

FIG. 3.11 A damaged segment of a concrete bridge, more details of the damage have been shown in the left picture. (Photo: Dr. R. Javaherdashti Collection.)

treatment, either use of biocides or even trying to materials selection.

Yet there is another example we want to share it with our readers: in one of our inspections, we found out that under a concrete bridge, there was a rather significant defect as shown in Fig. 3.11.

Although we do not want to flag the case shown in Fig. 3.11, as to be related to MIC, we would like to flag it as an example of Zugzwang effect where it is actually too late now to do anything to rehabilitate the structure. We can certainly try to "repair" the concrete cover, but the metallic part inside is already showing significant signs of corrosion. Based on the white-colored material seemingly oozing out from the concrete near the damaged area, we are probably looking at a case of "efflorescence"; we can deduce that moisture is being taken in into the concrete bulk structure, thus exposing the whole metallic bar inside the structure to the danger of corrosion. Fig. 3.12 shows microbial degradation of the composite material inside of a (wet) cooling tower in a power plant:

In a situation like the one shown in Fig. 3.12, as well as Fig. 3.11, we can easily see what we called as "Zugzwang effect." In fact, the main features common in these pictures (like many more examples of this kind) are that it really does not matter if you decide to do or not to do anything: if you decide to repair, it is a cost, and if you try to mitigate, it is of little or no use either.

FIG. 3.12 A picture from inside chamber of a power plant, more details have been shown in the bigger picture. The whole area of the composite material of the chamber had been covered with a greenish material that seemed to be algae. (Photo: Dr. R. Javaherdashti collection.)

Based on what we said, we can redefine Zugzwang effect in cases related to MIC with the following two characteristics:

1. For the structure/asset that has been exposed to MIC, thinking of any repair is not economical any more.

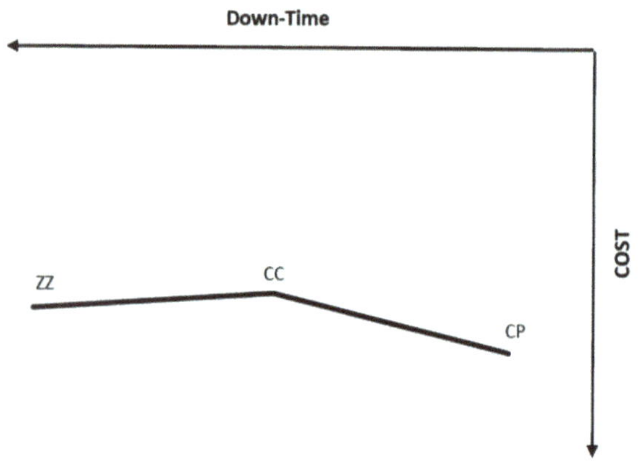

FIG. 3.13 Conceptual schematic of Zugzwang effect (ZZ), corrosion control (CC), and corrosion protection (CP).

2. Even if, despite its lack of economic feasibility, we decide to do it, MIC case has become so widespread and hard to cope with that it will make it technically not feasible to even try to mitigate MIC any more.

It is with this type of literature that **Zugzwang effect** can be taken as yet another terminology in corrosion management along with corrosion control and corrosion prevention (Fig. 3.13).

As it can be seen in Fig. 3.13, we show the relative position of **Zugzwang effect** in terms of cost and downtime involved with regard to corrosion control and corrosion prevention. As we mentioned in a previous footnote, corrosion control and corrosion prevention are quite different concepts. In addition to what we mentioned there, although corrosion prevention most of the time is applied to newly installed systems, corrosion control can be thought of being applicable to systems that are already in use and are facing with that kind of corrosion problems for which measures like changing the process parameters or making the system out of corrosion-resistant materials or anything of that sort is completely out of control.

What we have tried to show in Fig. 3.13 is that although the cost of corrosion prevention could be higher than corrosion control, the downtime factor involved will be smaller. In addition, the cost and downtime of a system that has been exposed to **Zugzwang effect** is so high that is incomparable with either of corrosion control and corrosion protection simply because this asset (the asset in which **Zugzwang effect** has occurred) is either out of service (the two fire water rings in the example we mentioned earlier) or is working under conditions that will require too much attention (in terms of operation and maintenance cost, constant likelihood of the asset stops working, and thus high risk of jeopardizing both production and safety issues).

In addition to the above, the figure—which is, by the way, conceptual that means it is not based on actual figures and, in accordance with TRIZ principles, is just a model to facilitate understanding the case—states that although the costs of corrosion prevention and corrosion control in some cases may be even close to each other, in terms of the downtime involved, it makes sense to prefer corrosion prevent to corrosion cost and both to **Zugzwang effect**. Besides, it can easily be seen that if in a given asset, the required monitoring, inspection, and treatment are not being carried regularly; this asset will be doomed to experience **Zugzwang effect.**

However, there is very important point here that we think needs clarification: during the normal life cycle of an asset, we try to manage the asset—at least from a corrosion management point of view—to withstand within the required, calculated, service life span. In other words, if a certain asset is expected to give reliable service (and thus be suitable for the job it is expected to do, the so-called FFS—fit for service) for, say, 20 years, and then it fails within much less time spans[e], then this is not a case of the asset experiencing its normal life cycle. This asset, if it has been severely damaged based on the two features we mentioned for **Zugzwang effect**, is highly likely to be replaced completely because obviously if it was curable by corrosion control, this

[e] "MIC has caused a lifetime reduction of flow lines in Western Australia from the designed +20 years to less than 3 years." See Cord-Ruwisch R. MIC in Hydrocarbon Transportation Systems. *Corrosion Australasia*. 1996;21(1):8–12.

asset would not have come to the end of its normal life so early.

It follows, then, that the main task of corrosion engineers is to avoid for the asset to arrive at experiencing **Zugzwang effect**. This way, by proper and constant monitoring and inspection of the asset, its present condition is known and this can be projected to its future condition based on a well-established corrosion management scheme. It is evident that by a corrosion prevention strategy, the time length to achieve **Zugzwang effect** will become much longer than just relying on a corrosion control strategy.

Any asset can go through three stages, FSS (fit for service), pseudo-FSS, and Zugzwang effect. All important and sensitive assets are to go through some procedures to make sure that they are safe to use. By applying checklists that essentially are the same in all industries, operators make sure that the equipment will "fit for its intended service," so the terminology FSS arises. At this stage of the life of the asset (or equivalently, this state of the service life of the asset), it can safely be assumed that there is no need for any further inspection as well as maintenance. At FSS state, the equipment is either new or quite recently tested or passed. It is also at this state of FSS that maximum reliability of the asset/equipment is assumed. Because of this critical significance, FFS is an essential part of any work procedure without which no work permit is to be issued.

When the part (or equipment) is permitted to be put into service, from the very moment it is in the service, corrosion will be at work on it. No matter if the part is new or recently inspected and repaired, corrosive factors will start operating on it quite extensively. At this state, the equipment cannot be said to be in its ideal FFS state anymore because, obviously, owing to corrosion the part is being depreciated. We call this state as "pseudo-FFS" in the sense that although the part is far from its ideal FFS state, it is still being used. Most of our operating assets are in their pseudo-FFS state.

There are two characteristics of pseudo-FFS state: service timewise, it is ahead of FFS and also that it requires both inspection and maintenance. These requirements for repair and maintenance as well as inspection are obvious because the part is being used and it is also exposed to corrosion (or a combination of corrosion with, say, erosion). This will decrease its reliability as the time goes by and it is only through regular inspections and, if required, repairs that the part can continue its service life.

However, as the part is more and more in service and its need for maintenance and inspection increases, it is possible that it will be left out. It is at this state that either there is no economical justification for repairs or repairs will be repeated over and over again because the root cause analysis of failures is not being investigated thoroughly and a good corrosion management treatment is absent. This is, as we have explained in required length above, the so-called "Zugzwang effect state." It must be noted that the Zugzwang effect state can be prevented by shortening the time of inspections (more inspections as the time for service life goes by) and applying root cause analyses should a failure occurs. This is a very important point to remember that although FFS and pseudo-FFS states are inevitable to arrive at during the service life of an asset, the Zugzwang effect state can be managed not to happen. Fig. 3.14 schematically shows the link between these three states of the equipment's service; life and its need for repair and maintenance, as well as inspection:

It may be somewhat confusing for some of our readers why we use asset, equipment, and component interchangeably. With very rough approximate definitions, each equipment is an asset, which has been composed of parts and components. A turbine is equipment that is also an asset for a plant. This turbine itself has been composed of many parts and components. However, each of these parts and components are themselves assets of the plant also, and if they fail, the whole turbine (equipment) is to fail. Because of this multifunctional nature that is being taken by each of these, we prefer to use them interchangeably.

Returning back to our original question of MIC and **Zugzwang effect,** obviously we want to avoid our assets to experience this effect. The practical

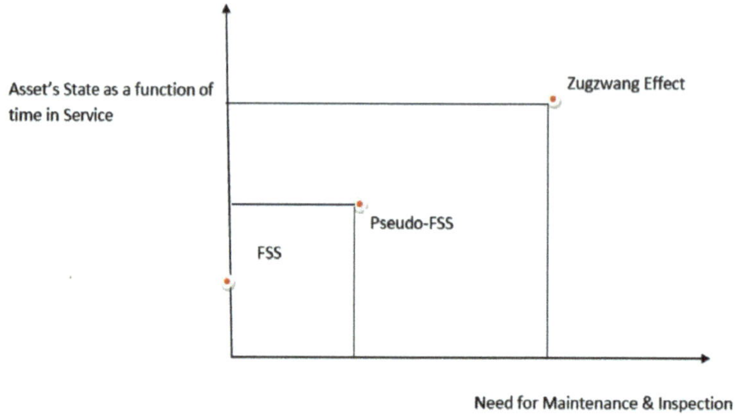

FIG. 3.14 The link between the three states of the asset's service life (FFS, pseudo-FFS and Zugzwang effect state) with the need that the component can have for repair/maintenance and inspection.

way we can propose is basically what we have in the chapter entitled "Diplomacy of MIC Treatment" in the new edition of our book.[15] In the conclusion section of that chapter, we reminded our readers of the following: ""… *In transferring from a* "Possible Future" *scenario to a* "Desired Future" *one, a manager has to take corrosion in his plant/company/workplace quite seriously. This can only be achievable through the application of both corrosion management (CM) and corrosion knowledge management (CKM) …. we [gave] more details on both aspects with a much more emphasis on CKM as applied to microbial corrosion cases (MCKM) in terms of four tasks that a strategist manager must undertake."* These tasks to be taken by a manager with regard to MCKM are as follows:

Task 1: "Appreciation of the need for "justification" of research in the field of MIC."

Task2: Appreciation of the need for establishment of a microbial corrosion research unit within existing R&D plan of the company.

Task 3. Organizing refreshing (applied) courses and workshops with regard to corrosion (in general) and microbial corrosion (in particular)

Task 4: Organizing the "human ware," (technical) employees and consultants who know how to deal with MIC efficiently and in a most feasible way.

All the four tasks mentioned above will make sense if corrosion and particularly MIC is seen as an integral part of management plan to be understood as important as HSE principles and to be applied with the same seriousness of improving productivity in an industrial unit.

Let us try to model what we mean by **Zugzwang effect** related to MIC by using one of the cases we mentioned above. We understand that this may be an imaginary scenario but we have good reasons to believe that when the seriousness of MIC was thoroughly and deeply appreciated by the company culture (that is, management), then even without knowing what **Zugzwang effect** is, it will be tried to avoid it.

Take the case we mentioned about the power plant with three fire water rings, two of which had been clogged. Fig. 3.15 shows the current and future situation of this plant:

As Fig. 3.15 shows, the current situation of the rings is that there are two cases with **Zugzwang effect** and one case in which either corrosion control or corrosion prevention measure must be applied. What is meant here by either corrosion control or corrosion prevention is that if the management decides, for example, to manage the leakage by applying some measures on the existing ring (for instance, use of internal coatings/linings), then it is a matter of corrosion control. However, if the management decides to entirely change the leaking ring and replace it with either a corrosion-resistant material (for instance,

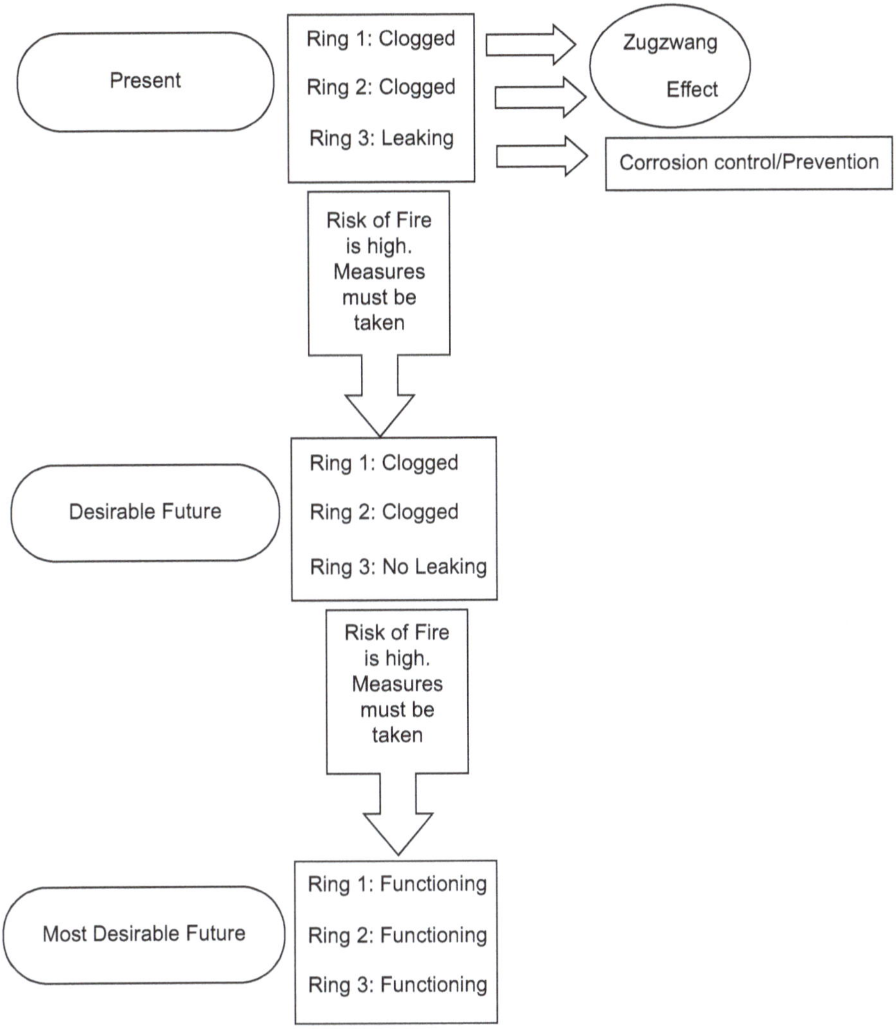

FIG. 3.15 Management of **Zugzwang effect** in a power plant whose three fire water rings have corrosion problems.

stainless steel) or again use the same material of carbon steel for the ring but this time apply measures to prevent MIC (for instance, continuous monitoring of the fire water chemistry and microbiology, preparing an MSL[f] [microbial species list], filtering the in-take fire water, keeping it moving even at certain intervals, etc..), this will be an example of corrosion prevention. No matter which measure for the third ring, the current situation is that to arrive at a desirable future; although there will be no particular change for the rings 1 and 2, ring 3 can be made to become leak-proof by either corrosion prevention or corrosion control measures.

As it can be seen, even in a desirable future, the **Zugzwang effect** will not change. The main reason, as we have several times emphasized upon, is that when an asset comes to its **Zugzwang effect** stage, it is basically hopeless.

[f]We have covered MSL in more detail in our upcoming book *"A Quick Reference to Failure Modes, Effects and Causes and Analysis (FMEA) of Microbial Corrosion (MIC)"*

However, we still have room for advancing the system to an even more desirable stage in the future, which is the most desirable future condition. In this third stage, in addition to finding a solution for the leaking ring, the two rings with **Zugzwang effect** have been rehabilitated and are now in a functioning state. Needless to say that if the rings 1 and 2 are now functioning, it is because they have undergone a corrosion prevention scheme. In other words, because their conditions were beyond any hope, then the whole asset has been upgraded/changed.

This can teach us that if the management decides to so-called "save" money on just applying corrosion control and leaving the assets with **Zugzwang effect** on their own, sooner or later they would have to pay even more than they seem to have saved for entirely changing the system through a corrosion prevention scheme. It is also evident, in our opinion, that as soon as based on a corrosion prevention scheme, the MIC scenario in a fire water ring is decided to be under control by a corrosion prevention scheme, the continuation will be based on a corrosion control scheme anyway.

Last but not least about **Zugzwang effect** and MIC is that it is through a four-task strategy of MCKM that moving from present to future and particularly desirable and most desirable futures will be possible. In other words, without a smart management of MIC cases through MCKM principles, the need for either upgrading from present situation to the future situations or avoiding to reach **Zugzwang effect** stage will not be felt and the end result is that the plant still works and perhaps doses have a good production rate, but owing to high rate of depreciation, which, in turn, is imposed by poor-planned corrosion management, the plant will not be in its desirable conditions.

We cannot close this section without mentioning what some industries may call it a smart way of dealing with corrosion: in industries where the level of awareness about corrosion is too low, in case there is a leakage or something, instead of long-term plans, it is preferred to apply short-term, so-called "smart" (!) plans. Most of the time, applying these plans can sustain the system seemingly working for a while and make the system go through an endurance period to try to do its expecting job. However, experience shows that these "smart" solutions are all doomed to shorten the time length for the asset to reach its **Zugzwang effect** stage. Two examples of such can be seen in Fig. 3.16.

It is of vital importance to note that the "smart problem shootings" of the kinds we have shown in Fig. 3.15 can only help bring the asset much quicker to its **Zugzwang effect** stage and must be seriously avoided.

"MYTHS EFFECT" IN RECOGNITION AND TREATMENT OF MIC

In previous section, we discussed the so-called "**Zugzwang effect**." An important factor that can lead into **Zugzwang effect** is what we call as "**myths effect**." In other words, if the former is a sign of being stuck in the present state of the system so that no further cure can do anything good anymore, the latter is a powerful tool that facilitates in making a system so vulnerable that whose end will be nothing but "**Zugzwang effect.**"

"**Myths effect**" is simply the wrong belief among technologists and professionals about MIC. The most destructive side of "**myths effect**" is that the person who has been affected by this effect not only has the power of ruining the assets and equipment by wrong recognition and treatment but also has the power of creating a huge obstacle in front of those who have done the job of recognition and treatment fairly good. The reason is that the professional diagnosed with "**myths effect**" syndrome has already wasted the resources so much that has also ruined the trust by his/her superiors on anything to be curable against MIC anymore.

Throughout our professional life, we have identified some examples of "**myths effect**" that we will share with our readers below:

1. SRB are the most important bacteria when it comes to MIC.
2. There is a relationship between the numbers of SRB and the rate of corrosion; thus the higher the numbers, the more severe the corrosion.

FIG. 3.16 Two examples of "smart" solutions to severe corrosion problems resulting in leakage. Use of metallic (above picture) and wooden (below picture) wedges! (Photo: Dr. R. Javaherdashti collection.)

3. If there are no sulfates in the environment, there is no need to look for SRB.
4. Applying an overvoltage of 100 mV on the CP potential will make the structure SRB-proof for sure.
5. Culturing is the best way to see if MIC can be an option in a corrosion scenario.
6. Polymers are more resistant to MIC than metals.
7. Silver has very powerful, long-lasting biocidal effects.
8. Stainless steels are more resistant to MIC.
9. Pit morphology can show if the corrosion case is microbially induced or not.
10. Copper and copper alloys are not affected by microorganisms.

In this section, we will just look at some of these myths:

Myth #1: SRB are the most important bacteria when it comes to MIC.

No, they are not. Reportedly,[39] it seems that the source of this 'SRB myth" has been started with W.A. Hamilton's work addressing MIC being "most commonly associated with sulfate-reducing bacteria." We have shown in previous pages that in addition to SRB, there are also other CRB that, if not more important than SRB, are at least as important. With regard to this myth, we should also add this fact that when biofilms are formed, it is highly likely to find mixed cultures of bacteria in which SRB can only be a certain fraction. Some experimental works with mixed cultures[40,41] have shown the corrosive effects of such mixed bacterial environments.

Myth #2: There is a relationship between the numbers of SRB and the rate of corrosion; the higher the numbers, the more severe the corrosion.

It seems that the source of this myth is the work by Ronay et al.[42] where it is stated that if the number of SRB per gram is less than 5×10^3, there is no risk of MIC, whereas a count of 10^4 or more of SRB per gram of soil is alarming a severe case of MIC. One of the ways by which CRB such as SRB can affect corrosion is by affecting the electrochemistry of the environment.[43]

Therefore, theoretically, even a few of them can alter the chemistry of the system to render it corrosive. In this sense, this is in contrast to what bacteria such as sulfuric acid–producing sulfur-oxidizing bacteria (SOB) do. It is obvious that the more the SOB, the more acid will be produced and so the corrosive effect will become more and more severe: a study[44] has shown that microbial sulfate oxidation was occurring and that the rates of this activity could be enhanced by increasing the population of sulfur-oxidizing bacteria in the samples.

Although research shows that there is no relationship between the number of SRB and corrosion rate,[45,46] these numbers can be useful when it comes to monitoring the system. Two examples of such situations are (1) indicating the possibility of getting more sessile bacteria and (2) the effect of biocide application in a closed system containing stagnant water. Therefore, the reduction in the number of planktonic bacteria, measured by most of quick tests, shows either they have been killed and no more swimming around or they have been transformed to sessile bacteria.[47]

Myth #3: If there are no sulfates in the environment, there is no need to look for SRB.

The name "sulfate reducer" may imply that first thing one should look for its existence in an environment to prove SRB presence will be sulfates as the bacteria need to reduce sulfate to sulfide. Research has shown that[48] in the absence of sulfate, SRB can switch to fermentation and ferment a variety of substances and produce hydrogen, carbon dioxide, and acetate. These products can then be used by other bacteria such as methane-producing bacteria (methanogenes) that by consuming hydrogen can accelerate corrosion through mechanisms such as cathodic depolarization.[49] Therefore, fermentative capabilities of SRB allow them to use alternative substances other than sulfate.

Myth #4: Applying an overvoltage of 100 mV on the CP potential will make the structure SRB-proof for sure.

The cathodic protection criterion of −0.95 V (vs. Cu-$CuSO_4$ reference electrode) to protect steel against SRB-induced MIC had its initial background in theoretical studies in 1964 by Hovarth and Novak[50,51] and, later, was experimentally verified by Fischer in the early 1980s.[30,52] Although the "−950 mV" criterion, or alternatively, 100 mV overvoltage criterion, has been widely applied in CP designs, we believe it is inconclusive. For example, in one particular case of cathodic protection, it has been reported that applying voltages up to−1.1 $V_{Cu\text{-}CuSO4}$ not only failed to prevent the growth of bacteria on the metal surfaces but also prompted the growth of certain microbial species and the rate of corrosion.[53] Also, the results of an investigation of CP effects on pure iron surfaces in the presence of SRB[54] have demonstrated that applying cathodic polarization of −1070 mV vs. Cu-$CuSO_4$ has not been sufficient to prevent the growth of SRB.

The results of a study on the effect of CP on the SRB biofilm formation on mild steel[55] showed that the efficiency of CP on SRB attachment on mild steel depended on factors such as the applied potential, type of SRB isolates, and the media. In our judgment, what these studies may suggest is that the 100 mV overvoltages must be treated as a flexible design guide and not a general law to which one needs to stick all the time.

Myth #5: Pit morphology can show if the corrosion case is microbially induced or not.

One of the ways by which recognition of the case of corrosion with MIC used to be advised was by looking at the shape of the pit.[56,57] As Tatnall and Pope put it "*surface morphology relates to the chemistry at the metal surface, not to the presence or absence of micro-organisms.*"[58]In fact, studies have shown that[59] the similarity between the dimensions of the bacterial cells that have adhered onto the surface and the size and dimensions of the pits may suggest that the pits have been initiated at the microbial attachment sites.

Myth #6: Copper and copper alloys are not affected by microorganisms.

Copper and copper alloys have had a reputation for being toxic to microorganisms, including, naturally, the CRB. However, there two important points to observe:

1. What is the effect of copper ions inside and outside the bacteria?
2. Copper resistance in bacteria.

There will be a huge difference if the copper ions (as caused by corrosion, for example) enter into the body of the bacteria than "hanging around." Fig. 3.17 schematically shows these two cases.

What happens when copper enters into the bacteria? There are many mechanisms through which copper may cause death of the bacteria. Some of these mechanisms are as follows[60]:

- Copper can interact with lipids, causing their peroxidation and opening holes in the cell membrane.
- Cu^{+2} can form protein chelates through a protein's carboxylate and amino groups, thereby inactivating a protein.
- Metals, such as copper, may disrupt enzyme structure.

Therefore, as long as copper is not introduced into the microorganism's vital systems, it is not doing anything. Alternatively, this may mean that copper and copper alloys are not immune to MIC till they find a way to be absorbed into the body of the bacteria. Experimental models for explaining microbial corrosion of copper have been explained in length elsewhere.[15]

In addition, there are documented cases where copper resistance behavior has been developed against copper. For example, Parvizi et al.[61] have reported of *Thiobacillus thiooxidans* being able to withstand copper ion (cuprous) concentrations as high as 20,000 ppm, Palanichamy et al.[62] have also observed endospore-forming genus *Bacillus* and non–endospore-forming genus *Propionibacterium* on copper surfaces. Critchley et al.[63] have reported the isolation of copper-resistant species such as *Sphingomonas* and *Acidovorax*. Needless to say that exact mechanism(s) of developing such copper tolerance still need more detailed investigations.

Therefore, copper under certain conditions can be also subjected to MIC, negating earlier "myths" that copper ions are toxic to all bacteria.[64] This can happen either through developing resistance mechanisms to copper or not being absorbed through the body of the bacteria. Also, the results of a study on the biological parameters related to MIC of copper[65] showed that microorganisms can protect themselves from the toxic copper ions by increasing the production of EPS (extracellular polymeric substance). Some researchers[66] believe that the bacteria and EPS together have a secondary role to the nature of the copper oxide film in controlling the rate of corrosion where the porosity of the oxide film seems to be the main corrosion rate controller.

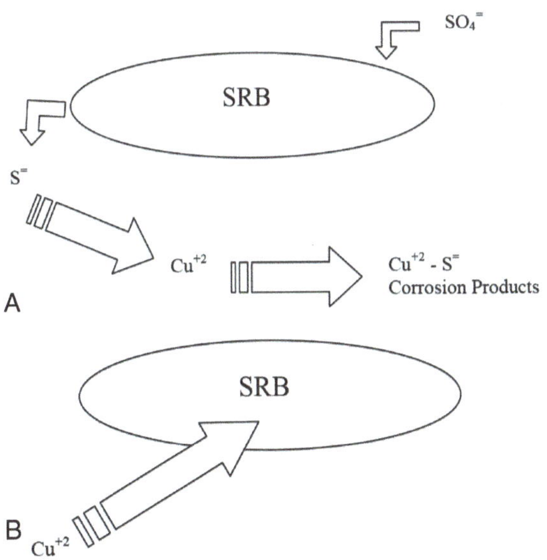

FIG. 3.17 **(A)** When copper ions combine with sulfides whose result is formation of the corrosion products. **(B)** When copper ions enter into the body of the bacteria whose result is the death of bacteria.

JAVAHERDASHTI MODEL FOR CM-MIC

Certainly, there are various models for corrosion management, from BP model to Total and Petronas'. These models are there to assist corrosion and integrity engineers to work out how to deal with corrosion more efficiently.

Here we would like to introduce our model, Javaherdashti corrosion management model, that can be applied into any particular corrosion case including MIC. We will first briefly introduce the model and then will try to address it by trying to place it for a corrosion case that will involve MIC.

However, before introducing our model, we have to explain what we specifically mean by a corrosion system. It is essential to exactly know the meaning behind a corrosion system because in many literature written on the concept of corrosion, authors assume that the readers already know what is meant by addressing a corrosion system. As we will see below, this is not that obvious a matter:

In a previous work,[67] Javaherdashti has fully explained the concept of corrosion system and its classifications. A corrosion system is defined as ***a part of universe in which corrosion occurs and is of interest to us.*** The two parts of this definition are of great significance: corrosion may occur, literally speaking, everywhere but not particularly of interest to us. In a refinery, for example, both underground fire water rings and metallic wires of the wire fence around the refinery area may be corroding. Whereas the former is of vital importance to us, the latter may not be taken that important. Therefore, not just because something is corroding but because its corrosion is important to us, we have to define and address a corrosion system properly.

Corrosion system can be assumed as to be consisting of subsystems such as A, B, C, etc. If corrosion problem of each subsystem "i" is shown as corr(i) and corrosion types observed in each subsystem as a,b,c, etc., corrosion of a system, or briefly, ***COFS***, is then defined as:

$$\mathrm{COFS} = \mathrm{Corr(A)} \cup \mathrm{Corr(B)} \cup \mathrm{Corr\ (C)}.$$

and

Corrosion in a system, or briefly, ***CINS***, is then defined as:

$$\mathrm{CINS} = \mathrm{Corr(i)} \cap \mathrm{COFS}.$$

Fig. 3.18 in the next section will show CINS and COFS schematically but at the moment it is necessary to note the physical meaning behind the mathematical expressions given above. COFS is a holistic approach toward the problem of corrosion in a system and most of the time, it is not feasible. In a boiler of a thermal power plant, there are many components from the waterwalls to the torches to the superheaters, etc. Although it is good to know ALL the corrosion problems in ALL of these components, practically speaking it is much feasible and much more useful if we adapt a CINS approach and try to locate each possible corrosion scenario for each component of a given boiler. We must not be confused by those who may be trying to reduce corrosion problems of a corrosion system to just a few. It is quite possible to concentrate on one particular corrosion mechanism and try to address it by a proper corrosion management approach, but this does not mean that other corrosion processes must be neglected. Nor does it mean that we have to forget about the possibility of series and parallel corrosion processes that may be expected. Therefore, it is a must to clearly understand our corrosion system, and then if we are talking about a multicomponent corrosion system, we consider a CINS approach.

Javaherdashti CM model consists of four main phases:

Phase 1, which is perhaps the most important phase as the rest of the phases are defined based on that, is called "definition." In this phase, it is important to define corrosion system as well as corrosion mechanism, whether or not these corrosion mechanisms are in series or in parallel with other corrosion processes that can be expected in the system and also possible corrosion measures that can be thought of for that particular corrosion system. During phase1, we have to be quite straight regarding the possibilities/uncertainties that can happen. At the beginning, we may have very few information that will allow us to define the corrosion system properly. It is a good practice to prepare a checklist of what we may consider as important parameters for definition of not only the corrosion system but also all available corrosion management measures. In this checklist, we must include all the factors in one way of another, which may affect the corrosion severity (in terms of corrosion rate) in the system of interest.

Phase 2 deals with application. When we could figure out the possible corrosion scenarios as well as anticorrosion measures in our defined corrosion systems, it is the time for application of the measure(s) we have considered the most feasible one(s). If for our given corrosion system and corrosion scenarios, we have decided that the best solutions are application of coatings; in phase 2, we have to apply the coating then. It is quite possible to decide on the type of coating required for this particular corrosion system based on factors such as, but not limited to, ease of application, economy, quality, and durability. Phase 2 is not just finished by application: one has to make sure that the selected corrosion management measure has been applied correctly and according to standards and procedures. If, for instance, we decide to apply CP in addition to coating, this is also another corrosion measure solution that needs to be designed and applied properly. To ensure that both solutions are intact in place, for coating we have to be careful that it does not have defects (holidays), whereas for CP we have to make sure that continuity exists in all sections and parts of the corrosion system.

Phase 3 concerns monitoring, which is itself a function of proper definition of KPIs (key performance indicators). Monitoring per se is not as important as having KPIs. The obvious reason is that we do monitoring to compare the present condition of the asset with what it was supposed to be. For example, according to NACE RP0775-2005, if the corrosion rate for general corrosion and localized corrosion is more than 10 MPY[g](0.254 mm/y) and 15 MPY (0.381 mm/y), respectively, corrosion is classified as severe. Likewise, the same recommended practice suggests that if corrosion rate for general corrosion is less than 1 MPY (0.0254 mm/y) and for localized corrosion is less than 5 MPYY (0.127 mm/y), corrosion can be taken as low. Now with a KPI for corrosion rates based on NACE RP0775-2005, it is possible to do monitoring: if the localized corrosion rates are more than 5 MPY, then the corrosion system is severely being corroded; however, if by applying proper corrosion measures we managed to lower the corrosion rate to almost 2 MPY (for localized corrosion), this would mean that our suggested corrosion management measures (phase 2) are working.

[g]MPY stands for mils per year. It is an alternative way of expressing corrosion rate.

Phase 4, usefulness, is second to phase 1 in importance. It consists of three performances: feedback, redesign, and information/experience sharing. The first action is related to whether the applied anticorrosion measure was good enough to use the obtained results to control corrosion in a given corrosion system. Redesign is how working conditions (and not necessarily the design of the system) can be modified based on new findings to minimize the likelihood of corrosion. This phase will also help a better, more practical corrosion risk management. The reason is that this phase comes after phases 2 and 3. Therefore, it is supposed to come up with solutions for corrosion in the system of interest. The third function to be performed during phase 4 is to share the good news, that is, to share the findings that helped us to overcome the corrosion problem. This is also a very critical step because many industries (especially in developing countries) lack the required legal/applicability framework that is necessary for sharing information. One of us (Javaherdashti) was once involved in a project for a well-known offshore oil and gas company in the Middle East. With a colleague, we found out that owing to existence of organic chlorine in the oil, many problems may arise, least of which being a sharp drop in the quality of produced offshore oil. Even presenting a review paper presentation about the mechanisms by which organic chlorine can be produced and without slightest mentioning of the name of that company prompted them to find our paper of "some intriguing" nature! Information sharing and privacy act are not related unless clearly defined. Without having a clear mind about what is to be said and what not, no information transparency can ever be expected. One of the results of such confusion is that owing to restricted flow of information, the very same mistake can be repeated over and over and every time with even much higher costs. Figs. 3.18 and 3.19 schematically shows Javaherdashti CM (corrosion management) model:

We would like to illustrate how Javaherdashti CM model can work in practice. Our corrosion system will be an underground fire water ring made up of carbon steel.

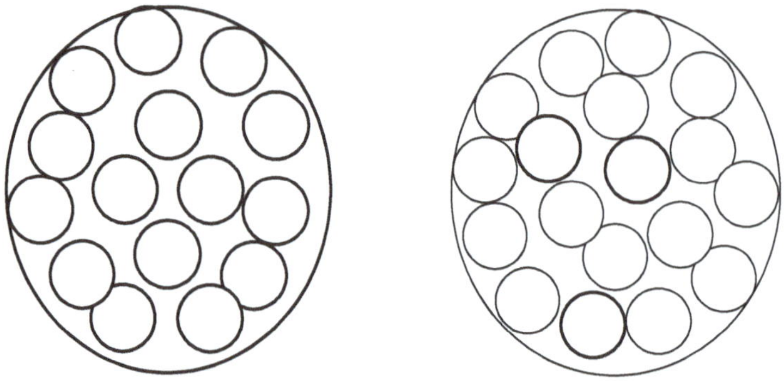

FIG. 3.18 Different approaches for considering the corrosion in a system: (left) COFS and (right) CINS approaches (the highlighted areas show subsystems whose corrosion is of interest to us).[67]

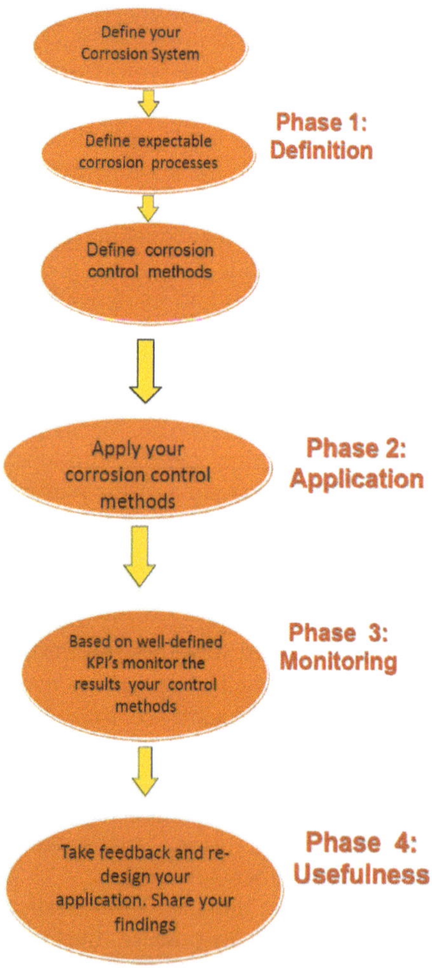

FIG. 3.19 Phases involved in Javaherdashti CM model.

Phase 1:

1-1: The ring is a simple system where a CINS approach will be applicable to it. The subsystems are the pipe itself (internal and external conditions), the fluid inside, the soil conditions, and the welding condition of the pipe. What could be the parameters that would allow us to define the fire water ring as our corrosion system? The material of the ring is very important: it is carbon steel but what type of carbon steel? There are varieties of carbon steels that may differ from each other in chemical composition and mechanical properties. Does the pipe (ring) have external/internal coatings? What are they? Have they been checked regularly? When was the last time that both the external and the internal conditions of the ring were checked? Is there cathodic protection applied on the ring or is it near installations that are under CP and thus it is likely that it will drain current? What is the relative layout of the ring with regard to the water table of the site? Is the underground ring (or a good part of it) below the water table level or above it? (If it is below water table level, it is as if the ring is in water.) What is the chemical/microbial composition of the soil around the ring and, say a meter, farther from it? What is the condition of welds? Are they continuous welding or spot welded? Is there a possibility for establishment of galvanic cells between the weld and the parent metal? Were the welds examined to be healthy and inclusion-free by routine methods such as radiography? If yes, what were the results(s)? Is the ring itself subjected to pre-commissioning practices (such as hydrotest)?

What is the flow rate of the water inside? Is it raw water or treated? If treated, how has been treated? What is the type of the water? Well water, river water, seawater? What are the physical, chemical, and microbiological conditions of the water? (For example, its conductivity, pH, temperature, flow rate, Langelier index, ...). Is it always at stagnation or from time to time it is circulated? How often it is circulated? Has there been any attempt to video-check inside the ring? Has there been any attempt to collect the debris from inside the ring and getting it analyzed from both chemical and microbiological aspects? If yes, what have been the results of such analyses?

1-2: It is important to have an idea about possible corrosion scenarios that can be expected and also if these corrosion scenarios are in series and/or parallel with each other. Stagnant water in a fire water ring could be taken similar to possible corrosion scenarios for stagnant hydrotest water. These scenarios have been discussed[10] and the references are given there. In short, the most probable corrosion scenario in such cases is MIC. This will give us an insight into what is required to look for and also what are the necessary measures to manage MIC.

1-3: What could be corrosion measure against MIC? There are a number of works we can do. Regarding the material of the ring, we can have a number of options, either upgrade the carbon steel from which the ring has been made to stainless steel (mind you, if we go for austenitic stainless steels 304 and/or 316/316 L, we have to be careful about the relation between chloride levels and temperature to avoid crevice and/or pitting corrosion of these stainless steels). Another alternative is to change from stainless steel to nonmetals such as composites or polymers such as HDPE (high-density polyethylene). This is despite new findings that both composites and polymers could also be vulnerable to MIC/MID.[15,68,69] Table 3.4 summarizes some of the factors that could be important on bacterial growth and activity.[15]

We can also have the alternative of internal coating of the carbon steel ring made from polymeric coating materials. Also it is possible to think of external surfaces of the ring exposed to soil made from MIC-resistant coating materials such as epoxy.[15] Both internal and external coating options can be made available based on economical considerations: is it feasible to change the material, or upgrade it or keep it as it is and apply both external and internal coatings? Yet, based on whether the ring is below water table level or above it, soil conductivity and moisture content, its oxygen content, its type (sand or clay sand with different soil particles sizes), we may consider different options for external coating as well.

Although the material and the soil around it are undoubtedly important factors, it may be a common practice to advise conditioning of the fire water ring: this can be made through two measures, which are physical-hydrodynamic and chemical. A good example of the former is to circulate the water from time to time not to let it become stagnant (flow rates below 5 m/s). The latter could be applying biocides.

TABLE 3.4
Some Environmental Factors Affecting Bacterial Communities

Effect of pH	Effect of DO (Dissolved Oxygen)	Effect of Ionic Strength
• Neutral pH favors bacterial growth • Low pH increases corrosion	• High DO increases corrosion • High DO will support the growth of GAB/APB assists, the survival of other bacteria (SRB and GAnB/APB) by symbiosis	• High ionic strength (conductivity), especially chloride, increases corrosion • High salt content slows bacteria growth • Lower salt content favors bacteria growth

APB, acid-producinge bacteria; *GAB*, generali aerobic bacteria; *GAnB*, generali anaerobic bacteria.

Circulating the water inside the ring may only be useful if inside the ring is clean and biofilm-free because it is quite highly likely that with making the fluid move, the nutrients necessary for bacterial establishments will also be shifted much closer to them, thus facilitating their growth (perhaps owing to the effect of fluid flow in enhancing mass transfer).[70] It is to be noted that although applying too high velocities may assist detachment of biofilm, it can introduce additional problems (e.g., erosion[h]).

Application of biocides can be another way to define possible corrosion countermeasures. Table 3.5 summarizes some of the most frequently used biocides with their pros and cons.

Application of biocides will be a very easy task from a practicality point of view. However, it is also a very risky business because, after all, biocides are generally synthetic chemicals and could have adverse, corrosive effects on the metallic substrates themselves.[71] It is always good to consider natural biocides as an alternative too.[72]

Phase 2:

Phase 2 can be regarded as being more practical with regard to the previous phase. It is in this phase that out of anticorrosion measures we have studied in phase 1, one or more will be applied. Suppose that based on the feasibility we did, we have decided to apply biocide treatment along with water circulation. Most of the time, these are easier and cheaper options than other options, but as our readers may have noticed, both of these solutions will only address the problem of internal corrosion, or more specifically, internal MIC issue and will not cover external corrosion, either in the form of MIC or else. This is also a very important factor that must be considered that for our particular corrosion system, is it the internal or external corrosion that matters to us the most. We have seen cases where failure of fire water has been due to external corrosion of the ring and no significant internal corrosion had been reported.

If we have decided to only apply biocide treatment and fluid flow turbulence in the form of circulation, there are few questions that we have to answer: for the part that biocide is to play, what biocide we are going to apply? Oxidising or nonoxidising (Table 3.5)? At what dosage and frequency the biocide(s) have to be applied? The reason we used plural form for biocide is that it may be decided that instead of applying one biocide type, two biocides to be used (dual regime). Nevertheless, three main parameters we have to define are the biocide type, the application regime, kill dosage, and required contact time. Table 3.6 gives some examples of biocides currently used in industries with their type, dosage, and required contact time to kill.

As we are dealing with living things, we would like to propose to do circulations in irregular periods so that each circulation will come as a shock treatment to the bacterial communities. It may be useful to try linear "detaching" velocities between 3 and 4 m/s (as applicable to potable water distribution systems[73]) when circulation is applied to make sure that biofilm growth is under control.

[h]In the context of biofilm engineering, erosion can have a relatively different meaning than what can come to minds of engineers: erosion is defined as "*The mechanical process of wearing or grinding objects away as particles of water wash over it*" (Holloway M, Nwaoha C. *Dictionary of Industrial Terms.* USA: Scrivener Publishing LLC- John Wiley and Sons, Inc.; 2013). However, when it comes to the interaction of generated shear forces in water and biofilms, at least two conditions can be imagined: "*biofilm deformation and frequent detachment of biofilm fragments (erosion), or even detachment of the entire biofilm (sloughing)*" (George E. Kapellos, Terpsichori S. Alexiou, Stavros Pavlou, "Fluid-Biofilm Interactions in Porous Media", https://www.researchgate.net/publication/280628377, doi:10.1016/B978-0-12-408077-5.00008-0, January 2015). Biofilm erosion is leading into a smooth biofilm surface, whereas sloughing will increase biofilm surface roughness (Kumarasamy MV, Maharaj PM. The Effect of biofilm growth on wall shear stress in drinking water PVC pipes. *Pol J Environ Studies.* 2015; 24(6):2479–2483). What we meant in the text was not biofilm erosion but the mechanical action of wearing the metallic substrate. Knowing this difference is very significant as erosion can refer to both the way the metallic substrate changes or the way the accumulation of bacterial film upon it does.

TABLE 3.5
Some Frequently Used Biocides and Their Pros and Cons[10,15]

CHLORINE
(+)
• Economical • Broad-spectrum activity • Effective • Monitoring dosages and residuals is simple
(−)
• Hazard concerns for the operator • Ineffective against biofilm bacteria • Ineffective at high pH • Inactivation by sunlight and aeration • Corrosive to some metals • Adverse effect on wood • Feeding (dosing) equipment is costly and requires extensive maintenance • Limitations imposed by environmental authorities on the discharge of chloramines and halomethanes
CHLORINATING COMPOUNDS (BLEACH [NAOCL], DRY CHLORINE [CA (OCL)$_2$])
(+)
• Circumvent the danger of handling chlorine • As effective as chlorine
(−)
• Can cause scaling problems • Expensive • Larger quantities needed than when using gaseous chlorine
CHLORINE DIOXIDE (ClO_2)
(+)
• pH insensitive • Good oxidizing agent for biomass • Tolerates high levels of organics • Dissolves iron sulfides
(−)
• Special equipment is required for generation and dosing • Toxic • Expensive
CHLORAMINES (LIKE AMMONIUM CHLORIDE)
(+)
• Good biofilm activity • Good persistence in long distribution systems • Has reduced corrosivity • Low toxicity
(−)
• Ammonia injection is required • Costs more than chlorine alone • Poor biocidal properties compared with free chlorine

Continued

TABLE 3.5
Some Frequently Used Biocides and Their Pros and Cons[10,15]—cont'd

BROMINE
(+)
• More effective than chlorine at higher pH • Broad-spectrum activity on bacteria and algae over a wider pH range than hypochlorous acid • Bromamines are environmentally less objectionable and less reactive with hydrocarbons etc., reducing the production of halomethane
(−)
• Similar to chlorine compounds • Expensive
OZONE
(+)
• A natural biocide, effective as a detachment agent against sessile bacteria on stainless steel surfaces • Resembles advantages of chlorine • Nonpolluting and harmless to aquatic organisms
(−)
• Like chlorine, it is affected by pH, temperature, organics, etc. • Its oxidizing effect does not resist throughout the system so ozone is used in small systems or specific sites within larger systems • Ozone must be generated on site, requiring investment for installation and running the equipment
SODIUM AND HYDROGEN PEROXIDES
(+)
• Used as a sanitizing agent • Have many advantages as ozone
(−)
• Requires high concentrations and extensive contact tune (to kill the microorganisms) • Cheaper and more safe than ozone • Careful use not to stimulate corrosion
ALDEHYDES
1. Formaldehyde (HCHO)
(+)
• Economical
(−)
• Suspected of being carcinogen • High dosages are required • Reacts with ammonia, hydrogen sulfide, and oxygen scavengers
2. Glutaraldehyde
(+)
• Broad-spectrum activity • Relatively insensitive to sulfide • Compatible with other chemicals • Tolerates soluble salts and water hardness

TABLE 3.5
Some Frequently Used Biocides and Their Pros and Cons[10,15]—cont'd

(−)
• It is deactivated by ammonia, primary amines, and oxygen scavengers
3. Acrolein
(+)
• Broad-spectrum activity • Penetrates deposits and dissolves sulfide constituents • In highly contaminated waters, it is generally more economical/cost-effective than chlorine • No particular environmental hazards
(−)
• Difficult to handle • Reactive with polymers, scavengers, and violently reacting with strong acids and alkalis • Potentially flammable • Highly toxic to humans
AMINE-TYPE COMPOUNDS
1. Quaternary Amine Compounds
(+)
• Broad-spectrum activity • Good surfactancy • Persistence • Low reactivity with other chemicals
(−)
• Inactivated in banes • Foaming • Slow acting.
2. Amine and Diamine
(+)
• Broad-spectrum activity • Have some inhibition properties • Effective in sulfide-bearing waters
(−)
• React with other chemicals, particularly anions • Less effective in waters with high levels of suspended solids
HALOGENATED COMPOUNDS
1. Bronopol
(+)
• Broad-spectrum activity • Low human toxicity • Ability to degrade
(−)
• Available as a dry chemical • Breaks down in high pH.

Continued

TABLE 3.5
Some Frequently Used Biocides and Their Pros and Cons[10,15]—cont'd

2. DBNPA
(+)
• Broad-spectrum activity • Fast acting and effective (at a pH above 8, it must be used for quick kill situations) • No apparent difficulties related to effluent discharge with these materials when applied as recommended
(−)
• Expensive • Affected by sulfides. • Must be adequately dispersed to ensure effectiveness because of low solubility in water • Although effective against bacteria at low concentrations, higher concentrations are required to control most algae and fungi, making them less cost-effective
SULFUR COMPOUNDS
1. Isothiazolone
(+)
• Broad-spectrum activity • Compatible with brines • Good control of many aerobic and anaerobic bacteria (like antisessile bacteria) and have activity against many fungi and algae at acidic to slightly alkaline pHs • Low dosages are required • Degradable
(−)
• Cannot be used in sour systems • Expensive • Less cost effective when the system contains significant amounts of sessile or adhering biomass. In such cases, the use of a penetrant/biodispersant enhances the effectiveness of the biocide • Extreme care required because of potential adverse dermal effects; automated feeding systems are strongly recommended
2. Carbamates (Alkyl Thiocarbamates)
(+)
• Effective against SRB and spore formers • Effective in alkaline pH • Useful for polymer solutions
(−)
• High concentrations are required
3. Metronidazole (2-methyl-5 nitroimidazole-1-ethanol)
(+)
• Effective against SRB • Compatible with other chemicals
(−)
• It is specific to anaerobic organisms

TABLE 3.5
Some Frequently Used Biocides and Their Pros and Cons[10,15]—cont'd

QUATERNARY PHOSPHONIUM SALTS (QUATS)

(+)

- Broad spectrum of killing activity and good stability. They are generally most effective against algae and bacteria at neutral to alkaline pH.
- Low toxicity
- Stable and unaffected by sulfides

(−)

- Not effective fungicides at any pH
- Their activity is mostly reduced by high chloride concentrations, by high concentrations of oil and other organic foulants, and by accumulations of sludge in the system
- Excessive overfeed of some types of quats may contribute to foaming problems especially in open recirculating systems with organic contaminations

TABLE 3.6
Some Examples of Industrial Biocides With Their Characteristics (the Brand Names are Erased Intentionally)

Biocide Brand Name	Biocide Type	Dosage and Frequency of Use
	Nonoxidising	1000 ppm based on barrel water produced per day.
	Nonoxidising	300–500 ppm fortnightly
	Oxidising type	300 ppm, batch treatment every 6 monthly
	Nonoxidising	1000 ppm per batching a. For pipeline usually batching every 2 months b. For other facility, batching is done when the SRB content is more than 10^3 cells/ml
	Nonoxidising	The target rate to apply is 500 ppm at fine filters inlet and booster pump outlet; the frequency on the treatment is once a week on each location with 3 or 4 days in between (i.e., day 1 fine filters and day 4 booster pumps)

Phase 3:

To be able to define future of the corrosion system, we need to have milestones that will show how well we have traveled so far. These milestones are alternatively called "key performance indicators" (KPIs). To get the best results, we have to know the values before and after applying a certain corrosion control measure and then compare it with the related KPI to see how well we have been doing. In other words, based on the results obtained from 1–3 and also phase 2, which shows the past and present of the corrosion system, we can compare it with the KPI to see the possibility of achieving the future. Let us illustrate this by an example.

Assume that we have decided to apply nonoxidizing biocide "A" with a concentration of 400 ppm fortnightly based on the values given in Table 3.6 to an already corroding metallic fire water ring. Let us also assume that from past experiences, we already know that the corrosion problem is MIC in nature. It is evident then that we have already been successful in phase 1 (definition) and now we have decided to apply an option seems more feasible to us, that is, biocide application. Furthermore, we can also assume that we have been experienced enough in MIC corrosion management to arrange for tabulated form

TABLE 3.7
A Guideline to Conduct phase 3 of Javaherdashti CM Model. The "Before" and "Now" Values Refer to Measures Taken Before Applying Biocides and After That and Are Imaginary Values. However, the Target Values Are Based on Operator's Manual[74](GAB: General Aerobic Bacteria, GAnB: General Anaerobic Bacteria)

SRB

SRB (Planktonic) Counts/ml			SRB (Sessile) Counts/cm²		
Before	Now	Target	Before	Now	Target
More than 10^4	Less than 3	Less than 1	More than 10^5	Less than 10^3	Less than 10^2

GAB

GAB (Planktonic) Counts/ml			GAB (Sessile) Counts/cm²		
Before	Now	Target	Before	Now	Target
Less than 10^4	Less than 10^4	Less than 10^4	Less than 10^4	Less than 10^4	Less than 10^2

GAnB

GAnB (Planktonic) Counts/cm²			GAnB (Sessile) Counts/ml		
Before	Now	Target	Before	Now	Target
More than 10^6	Less than 10^3	Less than 10^4	More than 10^4	Less than 10^3	Less than 10^2

of data as shown in Fig. 3.7. In other words, we know the number of planktonic and sessile bacterial community within the pipe (ring) before and after applying the biocide. If this can happen in reality, that is to say, if we could manage to get these values, this would mean that our MIC monitoring is near to perfect. Why neat perfect and not perfect? Because in the same way we do with monitoring non-MIC electrochemical corrosion by applying weight loss coupons to get a corrosion trend, we should have done the same here too: we should have had a trend of planktonic and sessile bacterial numbers before and after applying a certain dosage of a given biocide to see if the trend and not the absolute numbers approach the target values. Nevertheless, having a tabulated data form as in Table 3.7 is more than what one could ask for in many industries.

In a ring, when water stays still for some time, its oxygen content will sharply decrease[75] and therefore the environment becomes more feasible for anaerobic microorganisms. That may explain the relative higher abundance of GanB in comparison with GAB.

As it can be seen from the values given in Table 3.7 and comparing them with the target values, after applying the biocide, it seems that the number of general anaerobic bacteria (including SRB) shows a decline—while still above the target values—whereas the number of general aerobic bacteria shows no significant change with applying the biocide. Our interpretation from these figures is that most probably the applied biocide dosage has been more effective on GAnB (including SRB) than their aerobic counterparts.

Phase 4:

It is very important to share the experience and skills in such a way that the very same problem will not occur again. In addition, it is also necessary to use the feedback from the choices we made and the options we applied (phase 1–3) in such a way that the system will not go through the very same troubles again. This is what we have previously fully covered under "Use of Information" and "Transparency" in another work.[67] Here we are not going to repeat all those; however, we have to emphasize on a very significant point here, even briefly: transparency is not a technique, it is a culture. In contrast to it may seem at first sight, lack of transparency will not help the plant. It is always prudent to keep in mind that *"Transparency could further enhance the mass media and interpersonal communication of the CEOs' communication strategy and improve the oil company's image."*[76] It is obvious that this is not true only for CEOs and particularly oil CEOs. If in your plant, you still have no policies and regulations that will allow you to share the information you have gathered from applying your skills and methodology that we described in previous sections, it is time to seriously consider how to address this defect. Without this, the CM model will not be completed and will be reduced to a series of data and information: it is obvious that in this regard, there will be no benefit for anyone.

CLOSING REMARKS

MIC is of no belittling significance in industry, nor is it a simple process. Because of the interaction between living and nonliving agents that occur during biocorrosion processes, a lot of complexity and uncertainties will occur and involved in this corrosion phenomenon (which is, by the way, not a corrosion type per se).

Being electrochemical in nature, MIC can be treated by using the four categories of MIC treatment that we mentioned before. However, a very new aspect of this treatment will be based on the possibility of using bacteria against bacteria, or more generally speaking, biological treatment of MIC.

Biological treatment of MIC will be the topic of the chapters to be followed from now on in this book!

REFERENCES

1. Javaherdashti R, Nwaoha C, Tan H. *Corrosion and Materials in Oil and Gas Industries*. USA: CRC Press/Taylor & Francis; 2013.
2. Guo Z, Liu T, Frank Cheng Y, Guo N, Yin Y. Adhesion of *Bacillus subtilis* and *Pseudoalteromonas lipolytica* to steel in a seawater environment and their effects on corrosion. *Colloid Surf B: Biointerf*. 2017; 157:157–165.

3. Vu B, Chen M, Crawford RJ, Ivanova EP. Bacterial extracellular polysaccharides involved in biofilm formation. *Molecules*. 2009;14(7):2535–2554.
4. Costerton JW, Lewandowski Z, Caldwell DE, Korber DR, Lappin-Scott HM. Microbial biofilms. *Annu Rev Microbiol*. 1995;49:711–745.
5. Javaherdashti R. Corrosion and biofilm. In: Kanematsu H, Barry DM, eds. *Biofilm and Materials Science*. Springer; 2015.
6. Darwin A, Annadorai K, Heiderbach K. *Prevention of Corrosion in Carbon Steel Pipelines Containing Hydrotest Water-An Overview*. Paper No.10401. USA: NACE; 2010.
7. Spormann AM. Gliding motility in bacteria: insights from studies of *Myxococcus xanthus*. *Microbiol Mol Biol Rev*. 1999;63:621–641.
8. Beech I, Bergel A, Mollica A, Flemming H-C, Scotto V, Sand W. *Simple Methods for the Investigation of the Role of Biofilms in Corrosion, Brite Euram Thematic Network on MIC of Industrial Materials, Task Group 1, Biofilm Fundamentals, Brite Euram Thematic Network No. ERB BRRT-CT98-5084*. 2000.
9. Javaherdashti R. Behaviour of stainless steel 316L in a marine mixed culture containing sulphate reducing and iron reducing bacteria. *Corrosion Mater*. 2011;36(1):52–57.
10. Javaherdashti R, Akvan F. *Hydrostatic Testing, Corrosion, and Microbiologically Influenced Corrosion a Field Manual for Control and Prevention*. USA: CRC Press; 2017.
11. Rodríguez JJS, Hernández FJS, González JEG. Comparative study of the behaviour of AISI 304 SS in a natural seawater hopper, in sterile media and with SRB using electrochemical techniques and SEM. *Corrosion Sci*. 2006;48:1265–1278.
12. Liu W, Zhao Y, Lu M. Corrosion electrochemical characteristics of X60 pipeline steel in SRB and CO_2 coexistence environment. *Acta Phy Chem Sinica*. 2008;24:393–399.
13. Lee AK, Newman DK. Microbial iron respiration: impacts on corrosion processes. *Appl Microbiol Biotechnol*. 2003;62(2-3):134–139.
14. Newman RC, Rumash K, Webster BJ. The effect of pre-corrosion on the corrosion rate of steel in natural solutions containing sulphide: relevance to microbially influenced corrosion. *Corrosion Sci*. 1992;33(12):1877–1884.
15. Javaherdashti R. *Microbiologically Influenced Corrosion-an Engineering Insight*. 2nd ed. UK: Springer-Verlag; 2017.
16. Al-Abbas FM, Kakpovbia AE, Olson DL, Mishra B, Spear JR. In: *Biodiversity Associated with Sweet Crude and Seawater Injection Systems: Implication for Microbiologically Influenced Corrosion", 15th Middle East Corrosion Conference & Exhibition, Paper no.14106*. Manama, Kingdom of Bahrain: BSE-NACE; February 2–5, 2014.
17. Javaherdashti R. A review of some characteristics of MIC caused by sulphate-reducing bacteria: past, present and future. *Anti-corrosion Methods Mater*. 1999;46(3).
18. Miller JDA, Tiller AK. *Microbial Aspects of Metallurgy*. NewYork: American Elsevier Pub. Co. Inc; 1970.
19. TPC-3. *The Role of Bacteria in the Corrosion of Oil Field Equipment*. National Association of Corrosion Engineers; 1982. Third printing.
20. Magot M, Ravot G, Campaignolle X, et al. *Dethiosulfovibrio peptidovorans* gen. nov., sp. nov., a new anaerobic, slightly halophilic, thiosulfate-reducing bacterium from corroding offshore oil wells. *Int J Syst Bacteriol*. 1997;47(3):818–824.
21. Javaherdashti R. *MIC and Cracking of Mild and Stainless Steels* (Ph.D. Thesis). Germany: VDM Publishing; 2010.
22. Tatnal RE. Introduction. In: Kobrin G, ed. *Microbiologically Influenced Corrosion*. Houston, TX: USA: National Association of Corrosion Engineers; 1993.
23. Bos P, Kuenen JG. Microbiology of sulfur-oxidizing bacteria. In: *Proceedings of Microbial Corrosion*. London: The Metals Society; 1983.
24. Sakaguchi T, Tsujimura N, Matsunaga T. A novel method for isolation of magnetic bacteria without magnetic collection using magnetotaxis. *J Microbiol Methods*. 1996;26:139–145.
25. Hughes MN, Poole PK. *Metals and Micro-organisms*. Section 5.9. New York: Chapman and Hall; 1989.
26. Blakemore RP, Frankel RB. Magnetic navigation in bacteria. *Sci Am*. 1991;245:42–49. December 1981.
27. Stott JFD. Assessment and control of microbially-induced corrosion. *Metals Mater*. 1988:224–229.
28. Bahaj AS, Campbell SA, Walsh FC, Stott JFD. The importance of environmental factors in microbially-influenced corrosion: Part 2. Magnetic field effects. In: *Microbial Corrosion, Proceedings of the 2nd EFC Workshop, Portugal*. 1991.
29. Minton NP, Clarke DJ, eds. *Clostridia, Biotechnology Handbook*. Vol. 3. USA: Springer; 1989.
30. Mansouri H, Alavi SA, Fotovat M. Microbial influenced corrosion of corten steel compared to carbon steel and stainless steel in oily waste water by *Pseudomonas aaeruginosa*. *JOM*. 2015;67(7). Springer.
31. Javaherdashti R. A Fuzzy approach to model RISK of MIC in a cathodically-protected pipe. *Anti-corrosion Methods Mater*. 2000;47(3).

32. Javaherdashti R. Modeling microbiologically influenced corrosion of N-80 carbon steel by Fuzzy calculus. *Metall Mater Trans*. 2004;35A. ASM-TMS.
33. Javaherdashti R, Nwaoha C, Ebenso EE. Fuzzy prediction of corrosion resistance of duplex stainless steel to biotic iron reducing bacteria and abiotic synthetic seawater environments: a phenomenological approach towards a multidisciplinary concept. *Int. J. Electrochem. Sci*. 2012;7:12573−12586.
34. Javaherdashti R. A background Fuzzy algorithm for biofilm formation. In: *MIC − an International Perspective Symposium. Extrin Corrosion Consultants, Curtin University, Perth, Australia 14−15 February 2007*. 2007.
35. Duddridge JE, Pritchard AM. Factors affecting the adhesion of bacteria to surfaces. In: *Proceedings of Microbial Corrosion, 8-10 March 1983*. London, UK: The Metals Society; 1983.
36. Caroll KC, Butel JS, Morse SA, Mietzner T. *Jawtz, Melnick & Adelberg's Medical Microbiology*. Section III. 27th ed. USA: McGraw Hill Education; 2016.
37. Kilbane J. In: *Effect of Sample Storage Conditions on Oilfield Microbiological Samples", Paper No. 3788, CORROSION 2014, Houston, TX, USA*. 2014.
38. Liu H, Frank Cheng Y. Microbial corrosion of X52 pipeline steel under soil with varied thicknesses soaked with a simulated soil solution containing sulfate-reducing bacteria and the associated galvanic coupling effect. *Electro Electrochem Acta*. 2018. accepted and in press.
39. Sanchez del Junco A, Moreno DA, Ranninger C, Ortega-Calvo JJ, Saiz-Jimenez C. Microbial induced corrosion of metallic antiquities and works of art: a crtical review. *Int Biodeter Biodegrad*. 1992;29: 367−375.
40. Lee AK, Buehler MG, Newman DK. Influence of a dual-species biofilms on the corrosion of mild steel. *Corrosion Sci*. 2006;48:165−178.
41. Javaherdashti R, Singh Raman RK, Panter C, Pereloma EV. Microbiologically assisted stress corrosion cracking of carbon steel in mixed and pure cultures of sulfate reducing bacteria. *Int Biodeter Biodegrad*. 2006;58:27−35.
42. Ronay D, Fesus I, Wolkober A. In: *"New Aspects in Research in Biocorrosion of Underground Structures", Corrosion' 87, Brighton, UK*. 1987.
43. Javaherdashti R. *Making Sense Out of Chaos: General Patterns of MIC of Carbon Seel and Biodegradation of Concrete*. Paper 016, Presented at. Australia: Australasian Corrosion Association; 2007. ACA 07.
44. Pryfogle PA. *Monitoring Biological Activity at Geothermal Power Plants*. USA: Idaho National Laboratory; September 2005. www.inl.gov/technicalpublications/Documents/3311002.pdf.
45. Ilhan-sungur E, Cansever N, Cotuk A. Microbial corrosion of galvanized steel by a freshwater strain of sulphate reducing bacteria (*desulfovibrio* sp.). *Corrosion Sci*. 2007;49(3):1097−1109.
46. Little B, Wagner P. Myths related to microbiologically influenced corrosion. *Mater Perform*. June 1997:40−44.
47. Javaherdashti R. Ten commandments to prevent microbiologically influenced corrosion in your system. *Mater Perform*. 2005;44(11).
48. Hansen TA. Carbon metabolism of sulfate-reducing bacteria". In: Odom JM, Sigleton Jr R, eds. *The Sulfate-reducing Bacteria: Contemporary Perspectives*. Springer-Verlag; 1993.
49. Tatnall RE. Introduction. In: Kobrin G, ed. *A Practical Manual on Microbiologically-influenced Corrosion*. Houston, TX, USA: NACE; 1993.
50. Kajiyama F, Okamura K. Evaluating cathodic protection reliability on steel pipes in microbially active soils. *Corrosion*. 1999;55(No.1):74−80.
51. Tiller AK. A review of the european research effort on microbial corrosion between 1950 and 1984. In: Dexter DC, ed. *Biologically Induced Corrosion*. Houston, TX, USA: NACE-8, NACE; 1986.
52. Fischer KP. *Cathodic Protection Criteria for Saline Mud Containing SRB at Ambient and Higher Temperatures*. Paper No.110, CORROSION/81. USA: NACE International; 1981.
53. Pope DH, Zintel TP, Aldrich H, Duquette D. Efficacy of biocides and corrosion inhibition in the control of microbiologically influenced corrosion". *Mater Perform*. 1990;29(12):49−55.
54. de Romero MF, Parra J, Ruiz R, et al. *Cathodic Polarisation Effects on Sessile SRB Growth and Iron Protection*. CORROSION 2006, Paper No.06526. USA: NACE International; 2006.
55. Watkins P, Cheung S, Beech I. The influence of cathodic protection on the attachment of sulphate-reducing bacteria to mild steel surfaces. In: Angell P, Borenstein SW, Buchanan RA, et al., eds. *Proceedings of International Conference on Microbially Influenced Corrosion*, May 8-10, 1995, New Orleans, Louisiana, USA. NACE International; 1995:12/1−12/13.
56. Jack TR. *Biological Corrosion Failures*. ASM International; March 2002.

57. Cubicciotti D, Licina GL. Electrochemical aspects of microbially induced corrosion. *Mater Perform.* 1990;29(1):72–75.
58. Tatnall RE, Pope DH. "Identification of MIC", chapter 8. In: Kobrin G, ed. *A Practical Manual on Microbiologically-influenced Corrosion.* Houston, TX, USA: NACE; 1993.
59. Little BJ, Lee JS, Ray RI. Diagnosing microbiologically influenced corrosion: a stste-of-the-art review. *Corrosion.* 2006;62(11):1006–1017.
60. Michels HT, Wilks SA, Noyce JO, Keevil CW, Copper alloys for human infectious disease control, In: *Materials Science and Technology Conference, Copper for the 21st Century, Symposium, September 25–28, 2005, Pittsburgh, PA, USA.*
61. Parvizi MS, Aladjem A, Castle JE. Behaviour of 90-10 cupronickel in sea water. *Int Mater Rev.* 1988; 33(4):169–200.
62. Palanichamy S, Maruthamuthu S, Manickam ST, Rajendran A. Microfouling of manganese-oxidizing bacteria in tuticorin harbour waters. *Current Science.* 2002;82(No.7):865–869.
63. Critchley M, Taylor R, O'Halloran R. Microbial contribution to blue water corrosion. *Mater Perform.* 2005;44(6):56–59.
64. Geesey GG, Bremer PJ, Fischer WR, Wagner D, Keevil CW, Angell P. "Unusual Types of Pitting Corrosion of Copper Tubes in Potable Water Systems" as quoted by Ref. 46.
65. Fischer WR, wagner D, Peinemann H, Arens P, Tuschewitzki GJ. The influence of corrosion on biological parameters relevant for microbially influenced corrosion of copper. In: Angell P, Borenstein SW, Buchanan RA, et al., eds. *The Proceedings of International Conference on Microbially Influenced Corrosion,* May 8-10, 1995, New Orleans, Louisiana, USA: NACE International; 1995:21/1–21/14.
66. Webster BJ, Werner SE, Wells DB, Bremer PJ. Microbiologically influenced corrosion of copper in potable water systems-pH effects". *Corrosion.* 2000;56(9):942–950.
67. Javaherdashti R, Nikraz H. *A Global Warning on Corrosions and Environment: A New Look at Existing Technical and Managerial Strategies and Tactics.* Germany: VDM; 2010.
68. Balasubramanian V, Natarajan K, Hemambika B, et al. High-density polyethylene (HDPE)-degrading potential bacteria from marine ecosystem of Gulf of Mannar, India. *Lett Appl Microbiol.* 2010;51: 205–211.
69. Guo J, Yuan S, Jiang W, Lv L, Liang B, Pehkonen SO. Polymers for combating biocorrosion. *Front Mater.* 2018;5:1–15. Article 10.
70. Wen J, Gu T, Nesic S. *Investigation of the Effects of Fluid Flow on Srb Biofilm.* Paper No. 07516, CORROSION 2007. TX, USA: NACE International; 2007.
71. Bartling C. Treatment, prevention of microbially induced corrosion. *Pipeline Gas J.* 2016;243(6). Last visited 09 September 2018. https://pgjonline.com/magazine/2016/june-2016-vol-243-no-6/features/treatment-prevention-of-microbially-induced-corrosion
72. Javaherdashti R, Akvan F. "Corrosion Management &Failure Modes, Effects, Causes and Analysis as Related to Microbiologically Influenced Corrosion (FMEA-MIC)". In press.
73. Cloete TE, Westaard D, van Vuuren SJ. Dynamic response of biofilm to pipe surface and fluid velocity. *Water Sci Technol.* 2003;47(5):57–59.
74. Olabisi O, Al-Shamari AR, Al-Sulaiman S, Jarragh A, Mathew A. *The Role of Bacteria Population Density in Wet and Dry Crude Asset Integrity.* Paper No. 5534, CORROSION 2015. TX, USA: NACE International; 2015.
75. Penkala JE, Fichter J, Ramachandran S. *Protection Against Microbiologically Influenced Corrosion by Effective Treatment and Monitoring During Hydrotest Shut-in.* Paper No. 10404. USA: NACE; 2010.
76. Lanham K. "CEOs in Crisis: An Oil Industry Perspective", unpublished work based on a Master's degree. As presented at: http://kipplanham.com/Portfolio.html, visited on 27/March/2010.

CHAPTER 4

An Introduction to Microbiology for Nonmicrobiologists

INTRODUCTION

Microbiologically influenced corrosion is a multidisciplinary approach, and those students, researchers, and specialists aiming for a career in this filed need some knowledge about microbial life and their interaction with the surrounding environments. For this purpose and with respect to the contents mentioned in "Chapter 5: Biologic Treatment of Corrosion," a brief introduction to the major area of microbiology is explained here.

CELL AND MACROMOLECULES

A cell is a basic structural and functional unit in all living organisms. The cell is composed of water, organic, and inorganic molecules. Most cells contain 70% or more water. Water is an essential component of a cell. It is required for numerous biochemical reactions such as respiration, nutrient transportation, waste product transportation, temperature regulation, etc. Inorganic molecules that constitute around 1% of cell mass are required for cell metabolism. Sodium, potassium, magnesium, calcium, phosphate, chloride, and bicarbonate are some of the vital inorganic ions in a cell. Organic molecules are the unique component of cells. There are four groups of organic molecules in a cell[1,2]:

Carbohydrates

Carbohydrates supply energy for the cell, attribute in cell structure, and act as a signaling molecule.

Proteins

After water, proteins are the most abundant components of a cell. Proteins do most of the work in the cell. They are structural components of cells. They also regulate cell functions. Proteins catalyze cell's reactions and transport molecules.

Lipids

Lipids store energy in cells. They act as signaling molecules and also as structural components of cell membranes.

Nucleic acids

Nucleic acids including deoxyribonucleic acid (DNA) and ribonucleic acid (RNA) carry cell information in their nucleotide sequence.

These four groups of molecules, which are called macromolecules, are responsible for distinct properties of living organisms. Each type of macromolecules is made up of its own building block. Amino acids are subunits of proteins, fatty acids are lipids subunits, simple sugars are building blocks of carbohydrates, and nucleotides are subunits of nucleic acids[3] (Table 4.1).

Amino Acids

Amino acids are made up of five essential elements: carbon, hydrogen, oxygen, nitrogen, and sulfur. Fig. 4.1 shows a general structure of amino acids. Amino acids are organic compounds containing one amino (−NH2) and one carboxyl (−COOH) functional group, one hydrogen atom, and one side chain (R) group, which is specific to each amino acid. Amino acids attach to one another by peptide bond (−CO−NH−), a covalent bond linking carbon number 1 of first amino acid to the nitrogen number 2 of second amino acids[1] (Fig. 4.2). Amino acid linkages result in the formation of peptide (which contains few amino acids) or proteins (containing hundreds of amino acids).

Generally, there are two types of protein molecules: globular proteins and fibrous proteins. Globular proteins are spherical proteins, which have functional roles in a cell. These types of

Biological Treatment of Microbial Corrosion. https://doi.org/10.1016/B978-0-12-816108-1.00004-5

TABLE 4.1
Brief Overview of Macromolecules

Macromolecule	Elements	Monomer	Polymer	Function
Carbohydrates	C,H,O	Simple sugars	Polysaccharide	Energy source, Short-term storage, Cell Building block, Signaling molecule
Lipids	C,H,O	Fatty acids	Lipid	Long term storage, Signaling molecules, structural components of cell membranes.
Proteins	C,H,O,N,S	Amino acids	Polypeptide	Various function such as structural component of cell, enzyme, messenger molecule, signaling molecule, Transporter molecule.
Nucleic acids	C,H,O,N,P	Nucleotides	Nucleic acid	Holds the genetic information

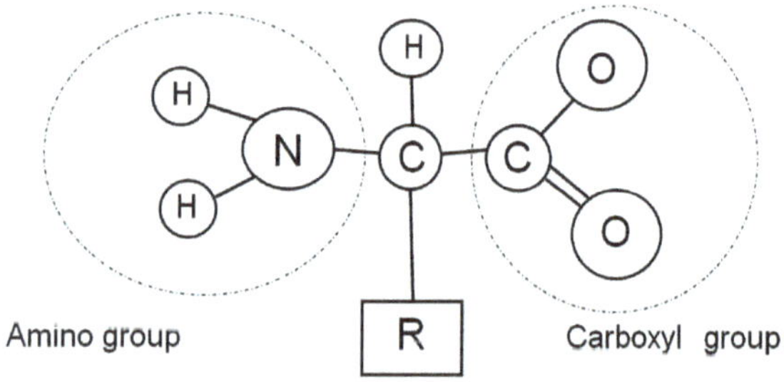

FIG. 4.1 Amino acid structure. Each amino acid molecule contains an amino group, carboxyl group, and a side chain or an R group.

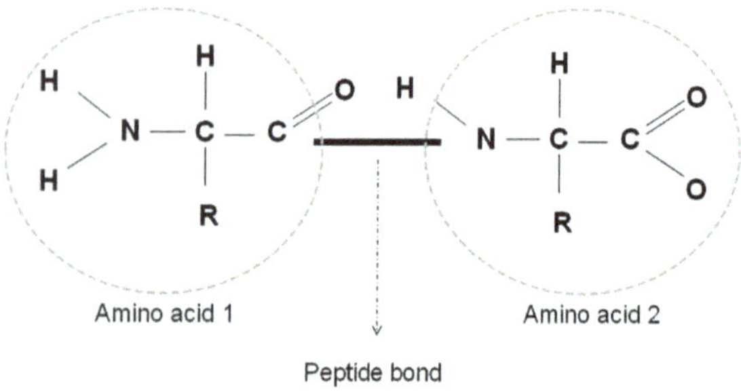

FIG. 4.2 Two amino acids are linked by peptide bond. Peptide bond forms when amino group of one amino acid links with the carboxyl group of the other, with the loss of a molecule of water.

proteins are soluble in water. In contrast to globular proteins, fibrous proteins have structural roles in a cell. Fibrous proteins are elongated and insoluble in water.[4]

Each type of protein, globular and fibrous, may have one or more of four types of protein structure[5]:

1. Primary structure
   The sequence of amino acids connected by peptide bonds forms the primary structure of a protein.
2. Secondary structure
   Secondary structure occurs when hydrogen bonding forms between a carbonyl group and an amino group of amino acid residues on the peptide backbone.
3. Tertiary structure
   The tertiary structure occurs due to interactions between side chains of amino acid residues. Several types of bonds and forces that hold a protein in its tertiary structure are hydrophobic interaction, hydrogen bonding, ionic bonds, disulfide bonds, Van der Waals forces, and salt bridges.
4. Quaternary structure
   This type of structure can be observed in the proteins made up of multiple polypeptide chains. Quaternary structure of these proteins forms due to noncovalent interactions between side chain groups of amino acids residue on separate chains.

As mentioned above, a single protein may have one or more of the protein structures.

Combination of these structures establishes the shape of protein, which finally determines the function of the protein. Therefore, any factor that affects the structure of a protein may manipulate the activity of the protein. For instance, it has been shown that increase in shear stress changes the tertiary structure of FimH (a protein mediates the adhesion of type 1 fimbriae of *Escherichia coli*) and improves the bacterial surface attachment.[6]

Fatty Acids

Fatty acids have consisted of hydrocarbon chain terminated with the carboxylic group (Fig 4.3A). The carboxylic head of fatty acid is polar and hydrophilic, whereas the long carbon chain of fatty acid is nonpolar and hydrophobic.[3] Generally, there are two types of fatty acids (Fig. 4.3B).

1. Saturated fatty acids that are having single carbon to carbon bonds in the hydrocarbon chain.
2. Unsaturated fatty acids that are at least having one double or triple carbon to carbon bond in its hydrocarbon tail.[1]

Fatty acids can also be classified on the basis of their tail length:

- Long fatty acids that have 16 or more numbers of carbon in their hydrocarbon chain.
- Medium length fatty acids that contain twelve carbons in their hydrocarbon chain.
- Short fatty acids that are having hydrocarbon chain containing 2–6 carbon atoms.

Fatty acids can make bonds with other molecules via their reactive carboxylic acid cap. For instance, phospholipids that are the most abundant type of lipid constitutes found in all cell membrane (the membrane that separates cell cytoplasm from surrounding environment) are composed of two fatty acids, one glycerol and one phosphate group.[3] During formation of phospholipids, the carboxyl group of each fatty acid reacts with one of the two carbon of glycerol molecule and the remaining carbon of glycerol links with phosphate group, which is usually connected to a polar molecule like choline (Fig. 4.4A).

Phospholipids can spontaneously form a lipid bilayer because of their amphiphilic (two fatty acids tail with hydrophobic character and a phosphate head with hydrophilic properties) nature (Fig. 4.4B). Fatty acids not only play a role in the formation of lipid bilayer structure of the membrane but also influence the membrane fluidity (or viscosity of cell membrane). Membrane fluidity that involves in the diffusion of biomolecule within the membrane is

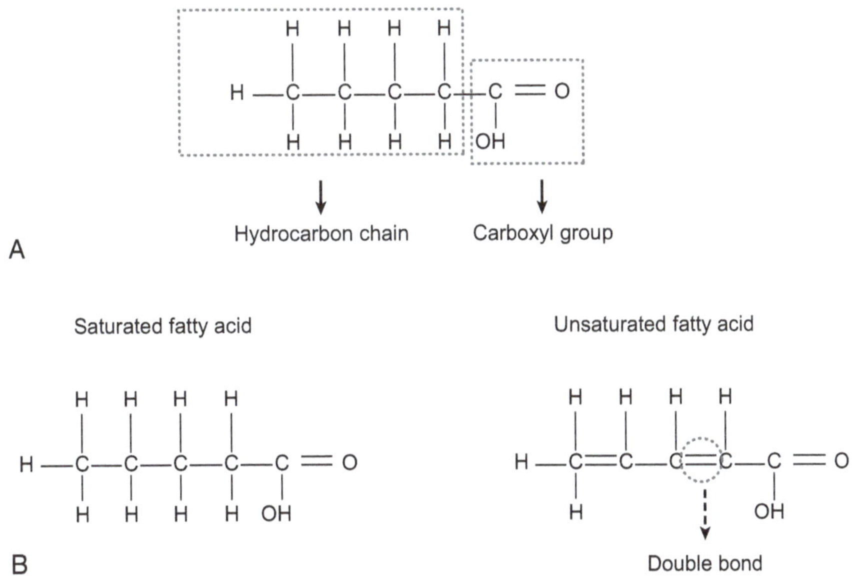

FIG. 4.3 **(A)** Fatty acid structure; each fatty acid is consisted of hydrocarbon chain with a carboxylic group on the end. **(B)** Chemical structure of saturated and unsaturated fatty acid.

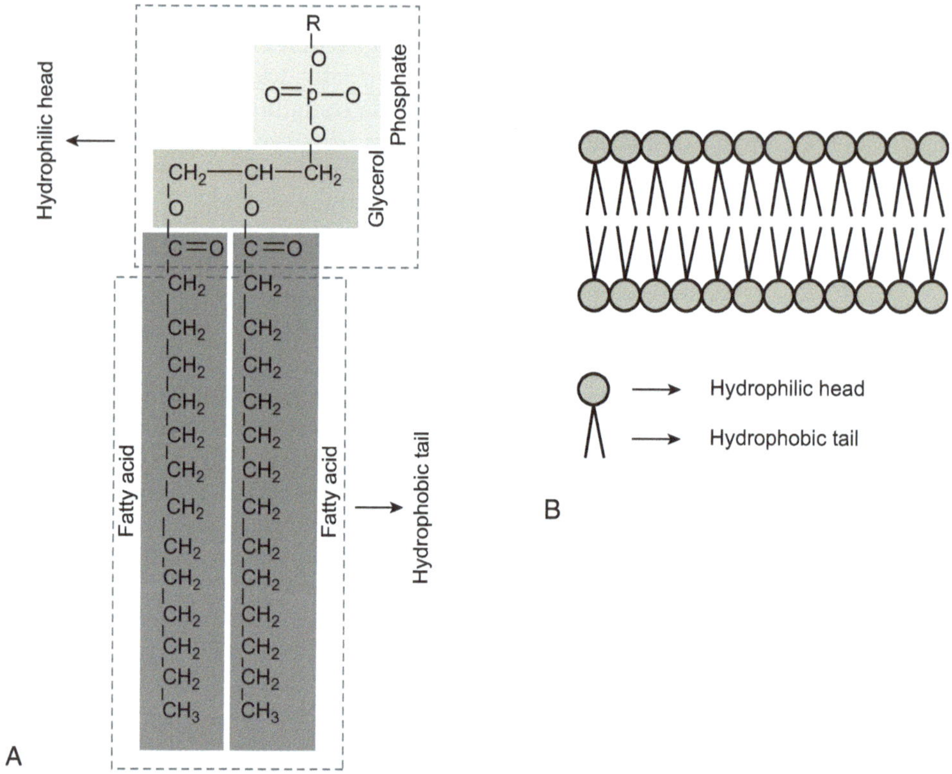

FIG. 4.4 **(A)** A phospholipid molecule is consisted of two fatty acids and a modified phosphate group, which is attached to a glycerol backbone. **(B)** Schematic representation of the arrangement of phospholipids in a lipid bilayer.

associated with the length and saturation of fatty acids (short and unsaturated one increase the fluidity of the membrane).[7]

Monosaccharides (or Simple Sugar)

Monosaccharides are carbohydrates with the general formula (CH2O). Each monosaccharide has consisted of several hydroxyl groups and either one carbonyl aldehyde or ketone group (Fig. 4.5). Monosaccharides with aldehyde group are called aldoses and those having ketone group are called as ketoses.[3] Monosaccharides can also be divided based on the number of carbon atoms they contain. For instance, monosaccharides with three, four, five, six, and seven carbon atoms are called, triose, tetrose, pentose, hexose, and heptose, respectively. Glucose, a hexose monosaccharide, is an energy source for many bacteria.

Monosaccharide can join together via glycosidic linkage and form disaccharide (chains of two monosaccharides unit), oligosaccharides (chains with few monosaccharide units), and polysaccharides (chains with many monosaccharide units)[1] (Fig. 4.6). Polysaccharides are an important component of bacterial biofilm. The polysaccharide can construct and maintain the biofilm structure. These macromolecules that make 1%–2% of biofilm matrix act as molecular glue and facilitate the bacterial attachment to the biotic or abiotic surfaces.[8] They also protect bacteria from various environmental threats such as physical stress and predation by phages or predatory bacteria.[9]

Nucleotide

Nucleotides are subunits of nucleic acids. They are made up of three parts: a sugar (the sugar part of deoxyribonucleic acids (DNAs) is deoxyribose

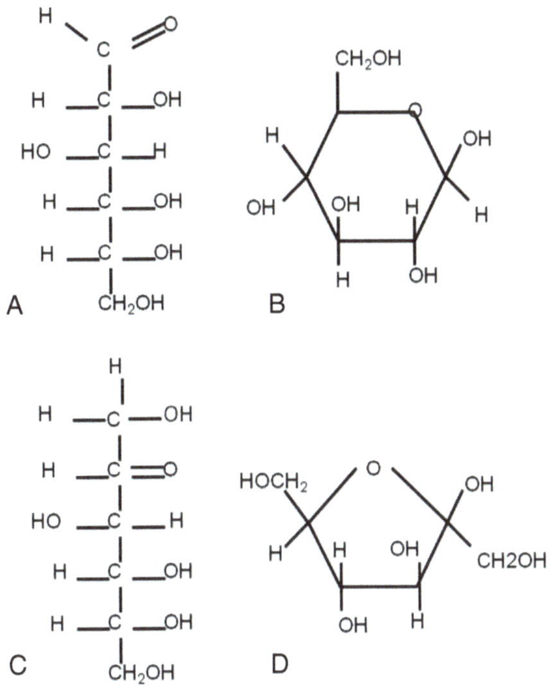

FIG. 4.5 **(A)** Long chain and **(B)** ring structure of aldose monosaccharide. **(C)** Long chain and **(D)** ring structure of ketose monosaccharide.

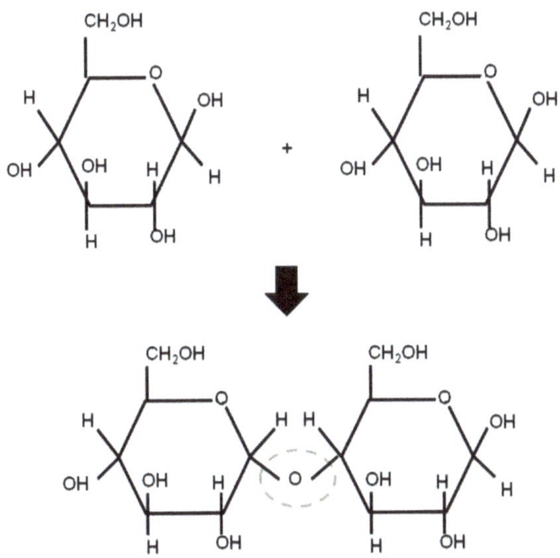

FIG. 4.6 Formation of glycosidic bond between two monosaccharide molecules.

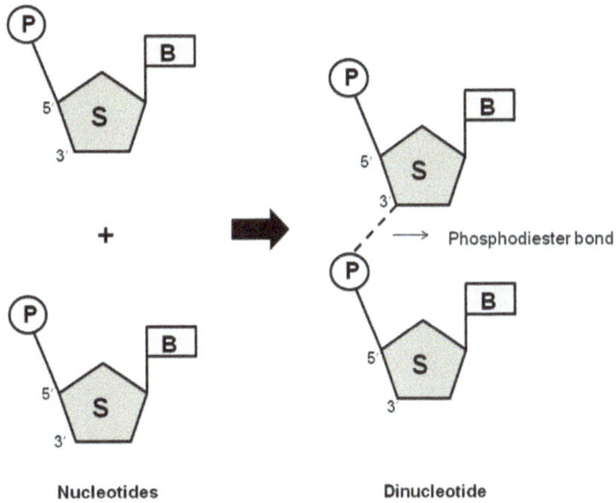

FIG. 4.7 Formation of phosphodiester bond between two nucleotides.

and in ribonucleic acids (RNAs), it is ribose), a phosphate group, and a base[10](Fig. 4.7).

Nucleotides differ from one another because of their bases. Each nucleotide may contain any of these nitrogenous bases: adenine (A), guanine (G), cytosine (C), and thymine (T). The bases used in both RNA and DNA are similar except in thymine. In RNA, thymine is replaced with uracil (U). Cytosine and thymine (uracil) are bases with a single ring and are called pyrimidines, and guanine and adenine are bases with double ring structure called purines.[3]

The nucleotides join together by means of a phosphodiester bond (Fig. 4.7). The phosphodiester bond links the phosphate group of one base to −OH group on the 3-carbon of the sugar of another base. Nucleotide joining results in the formation of oligonucleotides (strands of nucleic acids having a few nucleotides) or polynucleotides (chain of nucleotides with hundreds of nucleotides). A polynucleotide chain has no branches. Each strand of a polynucleotide is made up of an alternate sugar and phosphate group. The purine or pyrimidine bases join to a repeating sugar-phosphate group in the polynucleotide backbone by N-glycosidic bonds[10] (Fig. 4.8).

DNA is composed of two polynucleotide strands that are wounded around each other to give the double helix form. A close look at DNA shows that hydrogen bonds between nucleotide bases held the two strands together. In DNA a

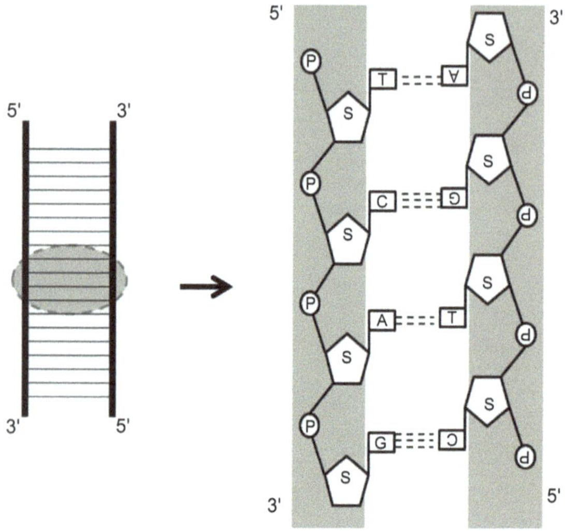

FIG. 4.8 DNA is double-stranded polynucleotide chain. Each chain is formed from sugar-phosphates backbone and bases linked to sugar moiety project from the backbone. The DNA base pairs are complementary to each other.

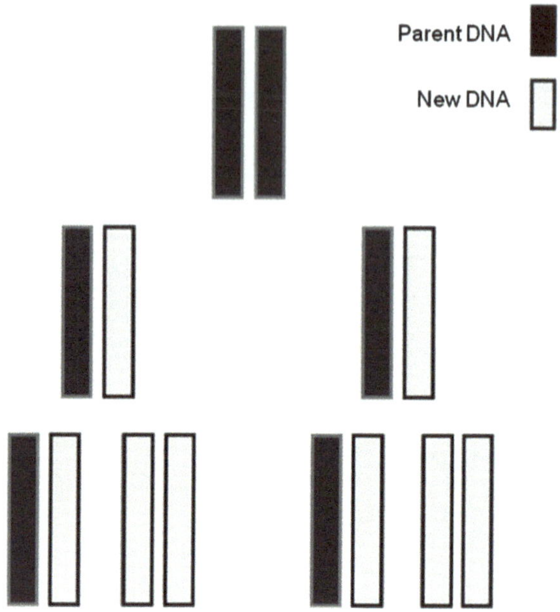

FIG. 4.9 Semiconservative model of DNA replication. In this model, each DNA contains one parent strand and one newly synthesized complementary strand.

nucleotide containing thymine will only pair with adenine and likewise, guanine always pairs with cytosine.[1] Because of the purine-pyrimidine pairing (adenine pairs with thymine only and cytosine pairs with guanine), each nucleotide pair always has three rings and therefore two strands of the helix are always evenly spaced. In addition, the pairing of bases in DNA gives rises into the formation of two complementary sequences in opposite direction; one $5' \rightarrow 3'$ and the other $3' \rightarrow 5'$.[3]

As mentioned previously, nucleotide sequences in DNA carry the cell information. This information can be inherited from mother cell to daughter cells in the process called semiconservative DNA replication (Fig. 4.9). During the DNA replication, two complementary strands of DNA helix are separated and each strand acts as a template for the production of a new complementary strand, with new nucleotides being added according to the rules of base pairing. Therefore, at the end of replication, each daughter cell comprises one parental strand and one newly synthesized strand.[10]

Opposite to DNA, RNA is a single polynucleotide strand. RNA has secondary and tertiary folding, which occur due to complementary base pairing within the same strand.[3] RNA molecules are involved in protein synthesis. There are four types of RNA[11,12]:

- Messenger RNA (mRNA)
  During a process called as transcription, DNA codon (sequence of three nucleotides that carry information for the synthesis of amino acids, rRNA or tRNA) is converted to mRNA. That means mRNA is a sequence of nucleotides complementary to the gene sequence of a DNA template (Fig 4.10).
- Transfer RNA (tRNA)
  tRNA acts as a bridge between the mRNA and the amino acid sequence of proteins (Fig. 4.11). At one end, tRNA binds to a specific amino acid and carries that to the protein synthetic machinery of a cell (ribosome). At another end, tRNA-linked amino acid makes a base pairing with the mRNA codon (a triplet of nucleotides in mRNA) via a complementary anticodon (nucleotide triplets in a specific part of tRNA) (Fig. 4.10).
- Ribosomal RNA (rRNA)
  rRNA is part of the structure of ribosome (Fig. 4.10). rRNA molecules act as a scaffold on

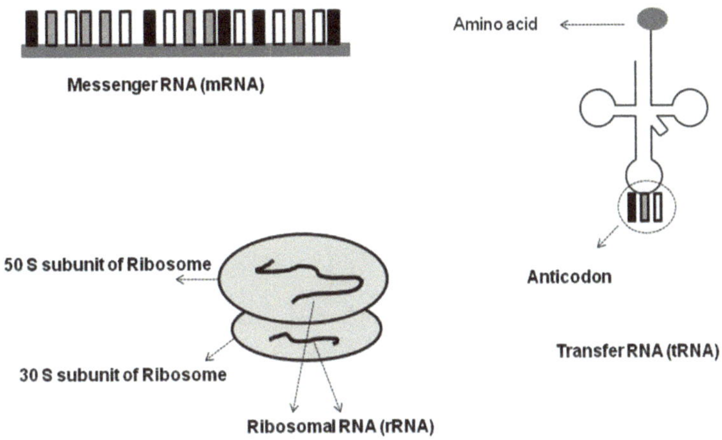

FIG. 4.10 Schematic representation of messenger RNA (mRNA), transfer RNA (tRNA), and ribosomal RNA (rRNA).

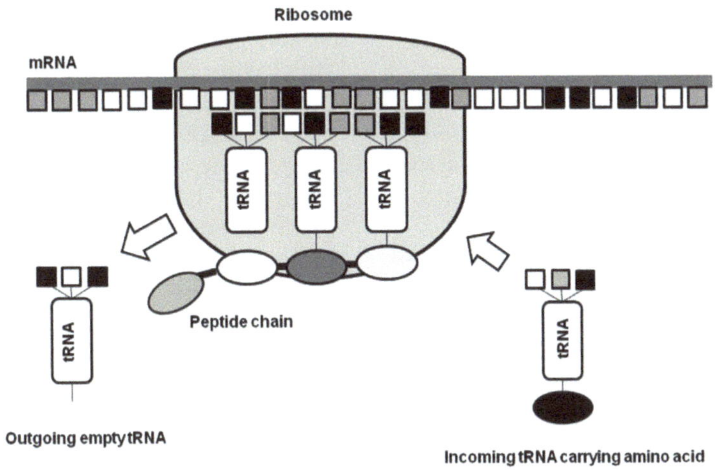

FIG. 4.11 During protein synthesis, tRNA molecules bind to specific amino acid and help that amino acids to join to the peptide chain by recognizing and binding it to proper mRNA codon.

which the ribosomal proteins can assemble. Small subunits of the prokaryotic ribosome (30S subunit) contain the 16s rRNA and large subunits (50S subunit) contain 5S rRNA and 23S rRNA. In addition to the structural task, these molecules have catalytic roles also. By recognizing and binding to a start codon in mRNA, 16s rRNA contributes in correct positioning of both the mRNA and the tRNA.

- Small RNA

  A small RNA is a short noncoding RNA that contributes in RNA silencing (a process in which RNA molecules inhibit gene expression or translation, by neutralizing targeted mRNA molecules). Small RNA plays several different regulatory roles in bacteria. It has been shown that this type of RNA helps bacteria to choose their lifestyle; whether to be a motile single bacterium or live as a multicellular surface-attached community.[11,13]

It should be noted that the living cells do not contain only molecules of the four groups outlined above. Some macromolecules consisted of more than one element, for example, lipopolysaccharides (carbohydrate and lipid), glycophospholipids (lipid, phosphate, and carbohydrate), and glycoproteins (protein and carbohydrate).

CLASSIFICATION OF ORGANISMS

To simplify the identification of one organism from another, biologists classify organisms into broad grouping, which share common traits. There are many different classification schemes. In the most accepted classification, which is called three domain-classification, both phenotypic (physical appearance of organisms) and genotypic (genetic properties of an organism) properties of organisms are considered. In this scheme, life is divided into three domains: bacteria, archaea, and eukaryotes.[14]

Each domain is further subdivided into:

- Kingdom
  - Phylum
    - Class
      - Order
        - Family
          - Genus
            - Species

In this classification, each organism is named by binominal nomenclature system; a system gives the scientific name to organisms by giving genus name, with the first letter capitalized, as the first name, and species name, all in a lowercase letter, as the second name. In the binominal system, the organism name should be underlined or italicized. For example, probably you have heard the name of corrosion-causing bacteria *Shewanella oneidensis*. This scientific designation shows that *Shewanella* is the bacterial genus and *oneidensis* is the species name.

Based on cell structures, all organisms in the three-domain system can be organized into two broad categories[2] (Fig. 4.12):

Prokaryotes

Prokaryotes are unicellular organisms lacking a nucleus or any other membrane-bound organelles within the cell. They are the most diverse and abundant groups of organisms on the earth. They are consisting of two major groups: bacteria and archaea.

Eukaryotes

Eukaryotes have cells with a nucleus and other organelles enclosed within membranes. Eukaryotic organisms are either unicellular or multicellular organisms. Table 4.2 shows the difference between prokaryotic and eukaryotic organisms.

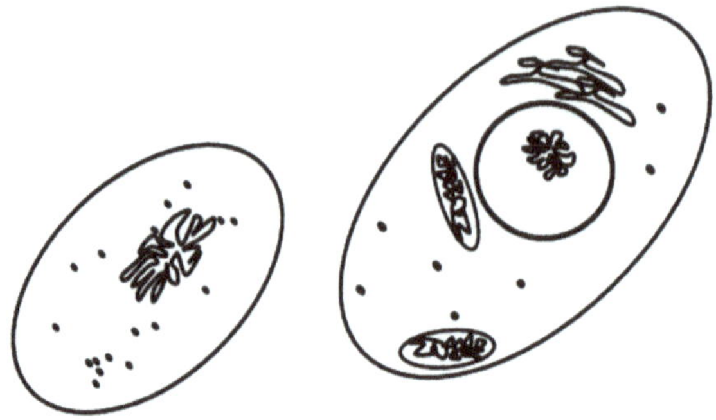

FIG. 4.12 Schematic representation of two basic cell types: prokaryotic vs. eukaryotic cells.

Tiny prokaryotic or eukaryotic organisms, which can be seen by microscope, are called microorganisms (Fig. 4.13). Bacteria and archaea are prokaryotic microorganisms and microscopic fungi; algae and protozoa are three groups of eukaryotic microorganisms.[14] There is another group of microorganisms, which is not classified in any of three domains, viruses.[14] These micro-organisms are lacking cell structure; therefore, they are neither eukaryotic nor prokaryotic. In this section, some general characteristic features of each kind of microorganisms will be discussed:

Bacteria

Bacteria are small organisms ranging from 1 to 3 μm and having a number of shapes, based on which they can be classified into spherical (cocci), rod (bacilli), spiral (spirilla), comma (vibrios), or corkscrew (spirochetes). Although bacteria are single-cell organisms, they can exist in pairs, chains, or clusters. They can be motile or nonmotile. They are reproduced by binary fission, budding, and fragmentation. In unfavorable conditions, some bacteria form endospore, dormant, and nonreproductive structures, which are extremely resistant to the harsh condition.[15]

Archaea

Archaea are a group of microorganisms that have many structural resemblances with bacteria, and because of these similarities,

TABLE 4.2
Major Differences Between Prokaryotic and Eukaryotic Organisms

Characteristic	Prokaryotes	Eukaryotes
Type of cell	Unicellular	Unicellular or multicellular
Cell size	0.1–5.0 μm	10–100 μm
Nucleus	Absent	Present
Presence of organelles (mitochondria, chloroplast, Golgi, endoplasmic reticulum)	Absent	Present
Ribosome	Smaller (70S)	Bigger (80S)
Cell wall	Usually present and chemically complex	Just present in fungi and plant and chemically simple
Glycocalyx	Present	Present in some cells that lack a cell wall
Flagella	Small with simple structure	Bigger with complex structure
Chromosome	One circular chromosome	Several linear chromosomes
Plasmid	Many have plasmid	Plasmid is not common
Cell division	Binary fission	Mitosis
Sexual reproduction	Absent	Present

originally they were thought to be bacteria. However, the reliable genetic analysis showed that they differ from bacteria. Archaea have a unique cell membrane, cell wall, and ribosome.[16] Table 4.3 summarizes the common differences between archaea and bacteria. These unique cell compositions not only enable them to resist against antibiotics but also facilitate their growth in hyperextreme conditions including[1]

- Absence of oxygen and low redox potential
- Low (<3) pH of the environment
- High (>90°C) temperature
- High (>10% NaCl) salinity

Fungi

Fungi are eukaryotic organisms with typical cell size between 5 and 20 μm. A fungal cell contains rigid cell wall made up of chitins. Fungi are mostly growing in low moisture and low pH environments. They lack chlorophyll (unlike algae) and therefore they must obtain food from their environment. Microscopic fungi exist in two forms[17]:

Yeast

Yeasts are single cell microscopic fungi that are reproduced by budding.

Mold

Molds are filamentous fungi made up of tubular threads called hyphae. Hyphae grow at their tips, proliferate, and form a network of threads called mycelium. Fungi embed their vegetative hyphae into solid wastes and soil, and then they secret digestive enzyme from their hyphal tips and break the organic matter into smaller molecules. Some branches of hyphae will not penetrate into surrounding materials and they grow into the air. These aerial hyphae may develop and differentiate to form structures bearing spore; a specialized structure adapted for dispersal and survival in harsh environmental conditions (Fig. 4.14). Molds play a very important role in many environmental cycles such as cycling of carbon, nitrogen, iron, and sulfur. They also attribute to the solubilization of insoluble phosphates and silicates. Molds are the most important group of fungi in biocorrosion of metallic materials. There are many reports on fungus-induced corrosion in the fuel tank, posttension structure, and underground power cable.[18–20]

Algae

Algae are eukaryotic microorganisms with typical cell size between 10 and 20 μm. Algae are photosynthetic organisms.[21] So like plants,

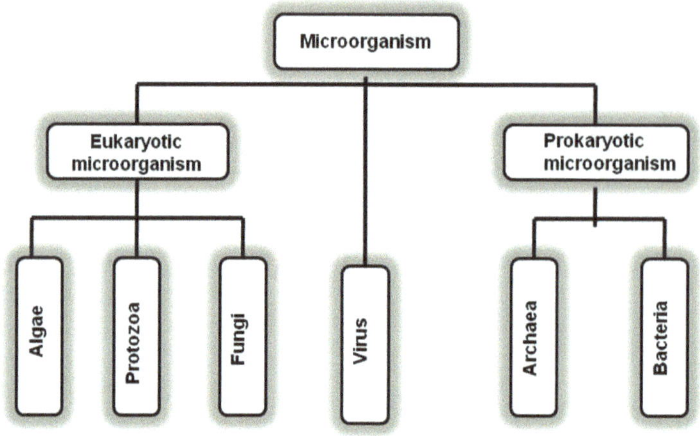

FIG. 4.13 Five different types of microorganisms.

TABLE 4.3
Major Differences Between Archaea and Bacteria

Characteristic	Bacteria	Archaea
Cell wall	Contains peptidoglycan	No peptidoglycan
Membrane lipids	Contain ester-linked, unbranched hydrocarbon	Contains ether-linked, sometimes branched hydrocarbon
Antibiotic sensitivity	Yes	No
DNA organization	Circular	Circular with histones
Plasmid	Common	Occasional
Found in	Everywhere	Extreme environments

they make their own food. Algae live either independently or in association with other organisms such as fungi (lichens). Most algae grow in an aquatic environment, and some live in a terrestrial environment where enough moisture is available. Algae are found among the fouling community, and they are known to be responsible for soft fouling.[22]

Protozoa

Protozoa are single-cell organisms with a typical cell size of 10–50 μm. Unlike fungi, first they absorb food and then they digest it inside the cell. They eat bacteria, other protozoa, and sometimes fungi as their food. Some protozoa are motile and some are not. Some protozoa are parasites and some are predators of bacteria, algae, and other protists.[16] Protozoa excrete nitrogen and phosphorus at high rates. The predatory activity of protozoa makes them an interesting agent for

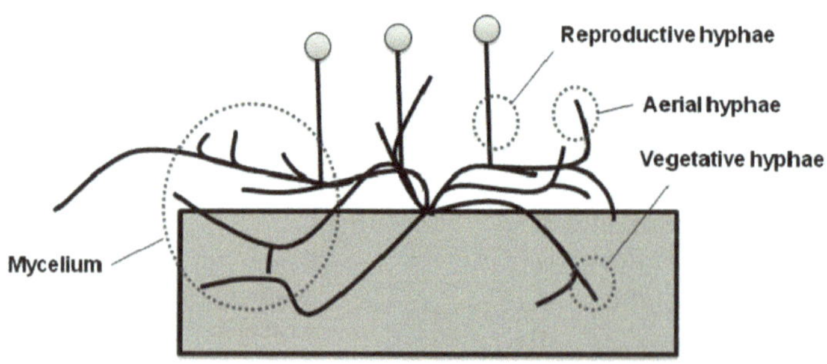

FIG. 4.14 Different types of hyphae in molds.

many industries. They are good potential for eradication of bacterial biofilm. By feeding on bacteria, they help in the production of good-quality effluents in the wastewater treatment system.[23] It has been shown that protozoan predation on oil-degrading bacteria influences the degradation of petroleum hydrocarbon.[24]

Viruses

Viruses are the smallest microorganisms with the approximate size of 10–300 (or 400) nm in diameter. A complete virus particle or virion is made up of genetic material (DNA or RNA) surrounded by a protein coat called capsid. Some viruses also possess additional membranous layer called envelope (Fig. 4.15). Viral envelope surrounds the nucleocapsid (capsid and genome combination).[16]

Viruses cannot replicate by themselves and they need host cell's replication machinery for their reproduction. Replication of viruses occurs in five steps (Fig. 4.16)[16]:

1. Adsorption—virus binds to the appropriate receptor of the host cell.
2. Penetration—virus injects its genome into the host cell.
3. Replication—copies of viral proteins and nucleic acids are replicated by using the host's cellular machinery.
4. Assembly—viral components assemble and full virions are developed.
5. Release—newly formed viral particles are released from the host cell.

Viruses can infect various types of organisms ranging from animals and plants to microorganisms. If viruses infect bacteria, they are called bacteriophage (phages). There are two types of bacteriophages called virulent phage and temperate phage. A virulent phage follows the lytic cycle of replication. In the lytic cycle, the progeny virions lyse the host cell at the end of replication[25] (Fig. 4.17A). A temperate phage undergoes the lysogenic cycle. In this type of life cycle, phage virions quiescently stay in a host cell (prophage) and replicate there without destroying the host cell (Fig. 4.17B). Progenies may pass on to subsequent generations or they may go through an induced (by exposure of host cell to stressful condition) lytic cycle.[25,26]

Recently, another route for phage propagation has been introduced. Under unfavorable growth conditions for the host cell (such as starvation), phages follow the pseudolysogeny. In this type of bacteria-phage interaction, phage nucleotide resides within starved host cell without multiplication (Fig. 4.17C). When growth conditions improved, development of temperate or lytic phages will initiate.[27]

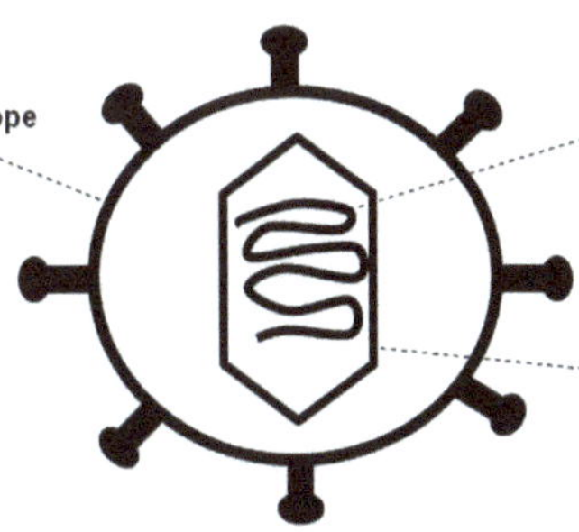

FIG. 4.15 General structure of viruses.

BACTERIAL CELL ARCHITECTURE

A variety of structures are found in the bacterial cell but not all of them are present in every genus. Structurally, there are three architectural regions in a bacterial cell:

1. Appendages
2. Cell envelope
3. Cytoplasmic region

Appendages

Appendages are attachments to the cell surface. Two types of appendages presented on the bacterial cell are flagella and pili (or fimbriae).

Flagella

Bacteria drive their locomotion with the help of lashlike structures called flagella (singular, flagellum) (Fig. 4.18). They may have one, two, or several flagella at one end or two ends of a cell. Each flagellum is composed of a long helical filament, a curved structure called hook, and a basal body consisting of a central rod and several rings. The basal body is embedded in the cell surface, whereas hook and helical filament are external to the cell.[28] The flagella beat in a propeller-like motion to help bacterium move

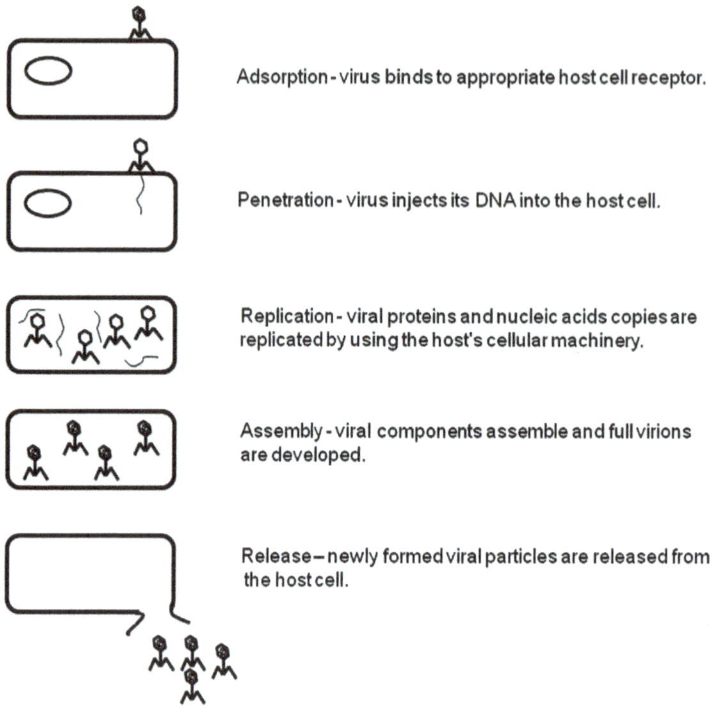

FIG. 4.16 Schematic representation of phage replication.

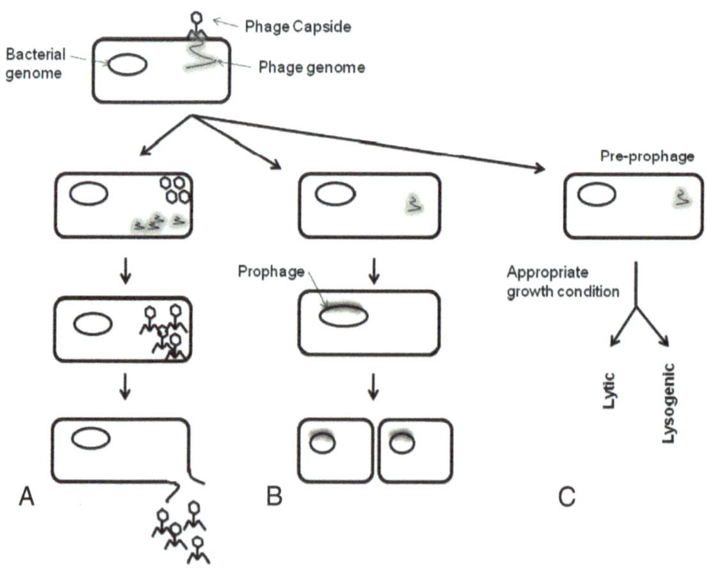

FIG. 4.17 Schematic representation of three different life cycles of bacteriophages: **(A)** lytic, **(B)** lysogenic, **(C)** pseudolysogenic cycle.

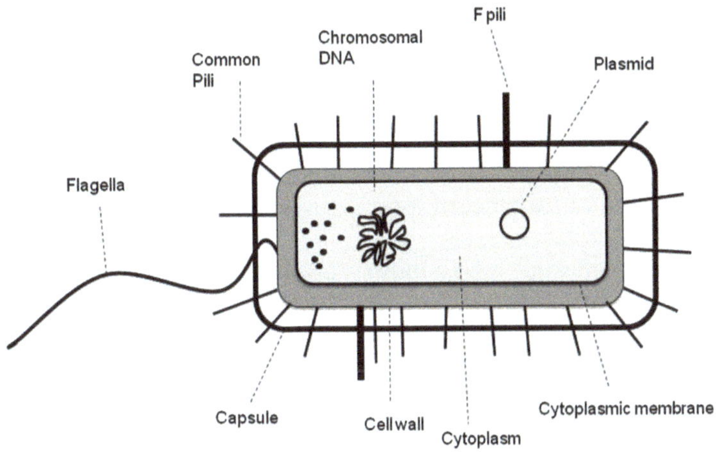

FIG. 4.18 Schematic representation of general structure of bacterial cell.

toward nutrients, away from toxic chemicals, or in the case of the photosynthetic bacteria toward the light.[29] Most of the flagellated bacteria have two modes of locomotion: swimming and swarming motility.

Pili

Pili (s., pilus) are small hairlike structures protruding from bacterial surface (Fig. 4.18). Pili are thinner, shorter, and more numerous than bacterial flagella and are primarily consisted of multiple proteinous subunits called pilin. Pili can be divided into two groups according to their morphologies, distributions, and functions[33]:

F pili (conjugation pili or sex pili): F pili facilitate bacterial attachment during the conjugation. This type of pili can act as receptors for bacteriophages. They are longer and thicker than the common pili (Fig. 4.18). Each bacterium can only have one to four F pili.

Common pili (attachment pili or fimbriae): Common pili that are distributed all over bacterial cell have several different functions (Fig. 4.18). They are responsible for irreversible bacterial adhesion on surfaces.[34] Twitching motility that is needed for bacterial migration along the surface and microcolony formation in young biofilms occur with the help of pili.[35,36] Pili not only are important for biofilm formation but also play a role in biofilm matrix stabilization.[37]

Some pili can act as electrically conductive nanowires. For instance, *Geobacter* nanowires are modified pili that help bacteria to directly transfer electrons from Fe(III) oxides.[38] Sherar et al.[39] observed that in media without organic electron donor, SRB produce nanowire-based pili by which they can directly harvest the electron from the carbon surface, promote cathodic reaction, and accelerate corrosion.

Cell Envelope

Bacterial cell envelope is made up of an interior cytoplasmic membrane, cell wall, and in some species of bacteria, an outer layer of glycocalyx.

Cytoplasmic membrane

Cytoplasmic (plasma) membrane is a microscopic layer, which encloses the bacterial cytoplasm (Fig. 4.18). This membrane is composed of a phospholipid bilayer with embedded proteins. The cytoplasmic membrane separates cell and protects it from the surrounding environment. It is a selective barrier that controls the transport of molecules in and out of the cell. Additionally, this membrane is a location for a variety of crucial metabolic processes such as respiration, photosynthesis, synthesis of lipids, and cell wall constituents.[16]

Cell wall

Cell wall that is located immediately outside of the cytoplasmic membrane is in charge of rigidity and overall strength of the bacterial cell (Fig. 4.18). It maintains cell shape, protects it from surrounding environments, and secures the cell appendages (pili and flagella).[40] A cell wall is mostly composed of unique bacterial substances called peptidoglycan (also known as murein). Peptidoglycan that is responsible for cell rigidity has consisted of sugars and amino acids.[41] Although all bacteria have peptidoglycan layers, the structure of peptidoglycan varies between species of bacteria.[40]

Some bacteria have thick and some have a thin layer of peptidoglycan. These two groups of bacteria can be distinguished by Gram staining. The former bacteria (Gram-positive bacteria) retain crystal violet dye and stain violet during Gram staining, while the latter (Gram-negative bacteria) do not; they easily lose the dye after washing. Table 4.4 summarizes the common differences between Gram-positive and Gram-negative bacteria. Gram-positive bacteria have a cell wall composed of 90% peptidoglycan and accessory molecules such as a teichoic acid (a glycopolymer) (Fig. 4.19). Teichoic acid has several crucial roles. It maintains the cell morphology, regulates the cell division, and generates the net negative charge of the bacterial cell.[42,43] In comparison to Gram-positive bacteria, the cell wall of Gram-negative bacteria is composed of an outer membrane that surrounds a thin layer of peptidoglycan (9%–10% of the total cell wall)[40] (Fig. 4.19). The outer membrane contains phospholipid, proteins, and mostly large molecules of lipopolysaccharide (LPS). LPS is responsible for the negative charge of the cell, the stability of the outer membrane, and protection of cell from toxic chemical and antibiotics.[44,45] In the cell wall of Gram-negative bacteria, there is a region called periplasmic space. This is the space between the cytoplasmic membrane and outer membrane of bacteria. Very active compartments of the cell such as enzymes for assembly of the cell wall and membrane component, various degradative or detoxifying enzymes, sensing proteins for chemotaxis, etc. are presented in this space.[46] Periplasmic space is also presented in Gram-positive bacteria, but it is much smaller than Gram-negative one.

TABLE 4.4
Differences Between Gram-Positive and Gram-Negative Bacteria

Characteristic	Gram-Positive Bacteria	Gram-Negative Bacteria
Peptidoglycan layer	Thick	Thin
Teichoic acid	Present	Absent
Outer membrane	Absent	Present
Periplasmic space	Absent	Present
Membrane permeability	More	Less
Endospore production	Present	Absent
Resistance to drying	High resistance	Low resistance
Resistance to antibiotic	More susceptible	More resistant
Type of toxin	Primarily exotoxins	Primarily endotoxins
Motility	Mostly nonmotile	Both motile and nonmotile
Reproduction	Binary fission	Binary fission, budding

A difference in structure and composition of cell wall influences behavior of bacterial cells in the environment. For instance, a thicker peptidoglycan layer in Gram-positive bacteria allows them to resist mechanical disruption. Presence of outer membrane that excludes penetration of antibiotics makes Gram-negative bacteria more resistant to antibiotic/biocides than their Gram-positive counterparts. Large periplasmic space in Gram-negative bacteria provides a temporary home for predatory *Bdellovibrio* bacteria and makes Gram-negative bacteria susceptible to bacterial predation.

Bacteria have response mechanisms by which they can modify cell wall components in response to stressful environments.[47] It has been shown that such changes in cell wall influence cell surface hydrophobicity, biofilm formation, as well as interactions with metal ions and antibacterial substances.[48–51]

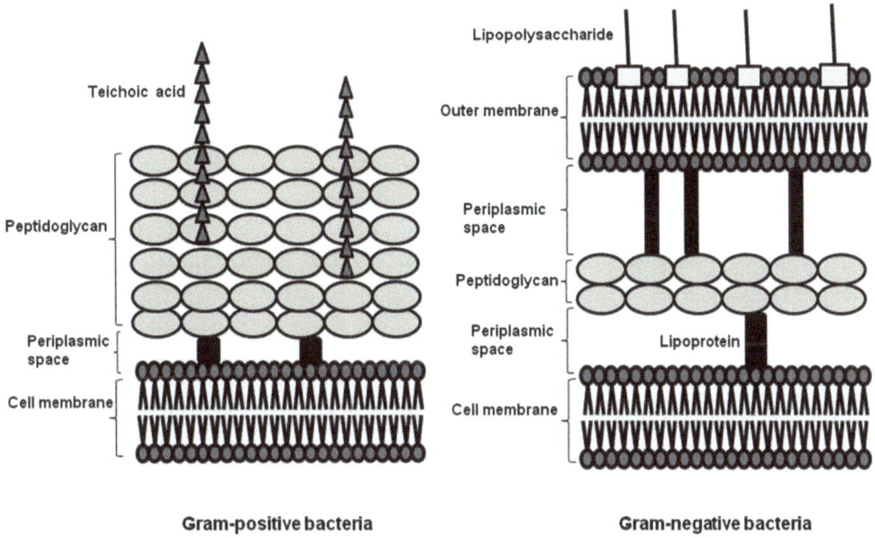

FIG. 4.19 Schematic representation of cell wall of Gram-positive bacteria and Gram-negative bacteria.

Glycocalyx

All cells have a cell membrane, most bacteria have a cell wall, and some have an additional layer of glycocalyx. A glycocalyx is a network of polysaccharide extending from the surface of bacteria. In addition to polysaccharides, they contain other materials such as protein and lipid. Glycocalyx layer protects bacteria from dehydration, loss of nutrient, and antibiotic.[52] This layer facilitates bacterial surface adhesion too.[53] Although the structural feature and chemical composition of the glycocalyx layer differ in various species of bacteria, generally there are two types of glycocalyx layer: capsule and slime layer.

Capsule: A glycocalyx considered a capsule when it is well organized and firmly attached to the cell wall[44] (Fig. 4.20A). It has been shown that encapsulated bacteria are more resistant to the bacteriophage infection.[45]

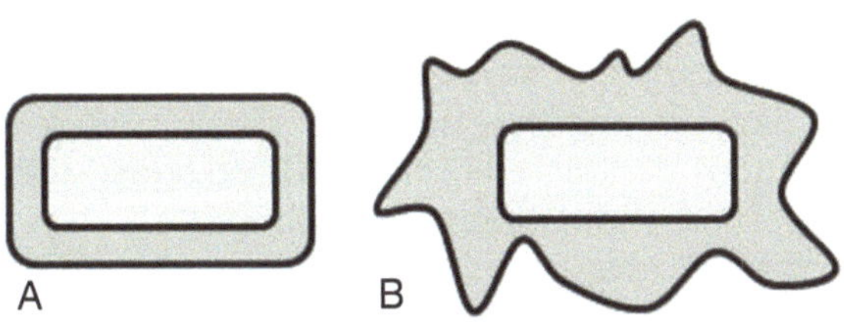

FIG. 4.20 Schematic representation of bacteria: **(A)** capsule and **(B)** slime layer.

Slime layer: In contrast to the capsule, slime layer is a zone of diffuse, unorganized material around the cell wall (Fig. 4.20B). It is loosely attached to the cell wall and can be easily removed from the cell.[54]

The slime layer should not be confused with S-layer. Similar to glycocalyx layer, S-layer is lying around the cell wall; however, as opposed to the glycocalyx, they are composed of a single molecular species of protein or glycoprotein in nature. Functional responsibilities of S-layers include the following[55]:

1. Bacterial protection against environmental threads such as ion and pH fluctuations, osmotic stress, enzymes, and predatory bacteria such as *Bdellovibrio*.
2. Maintenance of cell shape and cell envelope rigidity of at least some bacterial cells.
3. Promotion of bacterial surface adhesion.

Cytoplasmic Region

Cytoplasm is a thick fluid that fills each bacterial cell and is enclosed by a cell membrane (Fig. 4.18). Cytoplasm provides an aqueous environment for chemical reactions that take place in a cell. It is mostly composed of water, with some salts and proteins. It also contains some cell structures such as chromosome (DNA), ribosomes (biological protein synthesis machinery), and various sorts of inclusion bodies (granules of organic or inorganic material).[16]

MICROBIAL GENETIC MATERIAL

Each bacterium has a chromosomal DNA (Fig. 4.18). DNA carries all the instructions for bacterial growth, development, functioning, and reproduction. In a simple sentence, DNA is the cell's brain. Bacterial DNA may release into an environment by mechanisms related (or not related) to lysis of the cell. The released DNA which is called eDNA (environmental or extracellular DNA) is stabilizing the cell-to-cell adherence and biofilm architecture.[56–58]

In addition to chromosomal DNA, some bacteria have plasmids (Fig. 4.18). Plasmids are small double-stranded DNA. They are not necessary for bacterial growth and reproduction but give additional advantages to bacteria for survival. There are five types of plasmids[59]:

1. Plasmids that contain antibiotic resistance genes
2. Plasmids that carry bacteriocin; proteinous toxins that kill other bacteria
3. Plasmids that carry genes for production of F pili
4. Plasmids that turn nonpathogenic bacteria into pathogenic one
5. Plasmids that contain genes for the production of degradative enzymes.

Narenkumar et al.[60] studied the effect of *Bacillus thuringiensis* EN2 on the corrosion of mild steel, stainless steel, and brass in the presence and absence of plasmid inhibitors. They observed that the plasmid-cured strain was unable to form biofilm and it was less corrosive compared with untreated strain, which could form a biofilm.

Microbes are able to transfer their genes and attain characteristics that help them to adapt to a new environment.[61] There are three different mechanisms for bacterial gene transfer[2,61]:

1. Conjugation: In this process, a copy of plasmid transfers from a donor bacterium to recipient one through a sex pilus.
2. Transformation: In transformation, a recipient cell takes up DNA from the environment (eDNA) and incorporating it into its own genome. Hannan et al.[62] have shown that in the biofilm, the antibiotic-resistant gene can transfer from a donor bacterium, *Veillonella dispar*, to the recipient one, *Streptococcus mitis* by both conjugation and transformation.
3. Transduction: Transduction occurs when a bacteriophage acts as a vector and carries the gene from a donor bacterial cell to a recipient cell.

Because of minimum shear and closer cell to cell contact, the rate of lateral gene transfer is more at a high-density population of bacteria compared with low-density one.[63] Therefore, biofilm is an ideal environment for exchanges of plasmid genes.[64] It is interesting to know that the ability of biofilm formation is higher in strains of bacteria carrying conjugative plasmids.[65] By transferring the plasmid to recipient cells, donor bacteria promote the biofilm development in recipient on.

MICROBIAL METABOLISM

Microbes require energy to do work. To obtain energy, microbes must break chemical bonds of a substrate and form new bonds in the product. Much of the energy released from bond breaking and bond formation is dissipated in the form of heat, and remaining will be captured by cells in form of adenosine triphosphate (ATP), which is known as "molecular unit of currency" of intracellular energy transfer.[2]

ATP formation is coupled with the oxidation-reduction reactions in cells. During these reactions, an electron from high-energy substrate (electron donor) such as glucose transfers to an intermediate molecule called electron carrier. The electron carriers take the electrons to a group of proteins in the cell membrane called electron transport chain. As electrons move through the electron transport chain, they go from higher to lower energy level and ultimately pass to an electron acceptor molecule such as oxygen.[16]

Because microorganism can use different electron donors and electron acceptors, they live in a wide variety of environments. If an organism uses organic chemicals for its energy production, it is called as a chemoorganotroph, and if it uptakes inorganic one, it is known as chemolithotrophs. Below are some examples of chemolithotrophic microorganisms:

- Hydrogen oxidizers—these organisms oxidize hydrogen gas (H_2) with the use of a hydrogenase enzyme.[66]
- Sulfur oxidizers—these organisms are capable of oxidizing a wide variety of reduced and partially reduced sulfur compounds such as hydrogen sulfide (H_2S), elemental sulfur (S_0), thiosulfate ($S_2O_3^{2-}$), and sulfite (SO_3^{2-}) into sulfate (SO_4^{2-}). Sulfur oxidation occurs with the help of sulfite oxidase enzyme.[67]
- Nitrifying bacteria—nitrifying bacteria oxidize ammonia (NH_3) in a two-step process. In the first step, one group of bacteria called ammonia oxidizing bacteria convert ammonia to nitrite (NO_2^-) and in the second step, another group that is known as nitrite-oxidizing bacteria oxidize nitrite to nitrate (NO_3^-).[68]
- Iron oxidizers—these organisms obtain their energy by oxidizing dissolved ferrous iron (Fe^{2+}) to ferric iron (Fe^{3+}).

Chemolithotrophs and chemoorganotrophs are able to use a variety of electron acceptors. If they use oxygen as electron acceptor, they are called aerobic, and if they use electron acceptor other than oxygen, they are known as anaerobic microorganisms. Sulfate-reducing bacteria, nitrate-reducing bacteria, and iron-reducing bacteria, which, respectively, use sulfate, nitrate, and ferric iron as the final electron acceptors are some examples of anaerobic bacteria.[69–71] All of these anaerobic bacteria use the electron acceptors during dissimilatory pathways (i.e., they do not incorporate these compounds in the cellular constitutes [assimilatory pathway] but use them for energy production).

Microbes can produce energy from other pathways called fermentation. In fermentation that occurs in an anaerobic condition, bacteria use organic carbon as an electron donor, break it into products, and produce very little energy (production of 2ATP in comparison with 36 ATP molecules produced in aerobic respiration). Note that degradation of organic carbon occurs in absence of an exogenous or externally derived electron acceptor.[16]

REPRODUCTION OF MICROORGANISM

Microorganisms can reproduce by both sexual and asexual reproduction:

Sexual reproduction

In sexual reproduction, genomes of two parents are combined to create offspring with unique genetic profiles. Formation of genetic variation in sexually reproducing organisms increases the chance of survival against disease and environmental changes. Sexual reproduction is observed in some algae and fungi but, it is not occurring in bacteria and archaea.[72,73] Prokaryotes do not have sexual reproduction; however, to promote genetic variation, they exchange genetic information by horizontal gene transfer mechanisms such as conjugation, transformation, and transduction. Although the lateral gene transfer is sometimes called sex, note that this does not happen as a mean of reproduction. This process is just a survival strategy for prokaryotes.

Asexual reproduction

Bacteria increase their population through the asexual process. Asexual reproduction is a type of reproduction in which only one organism, without exchanging its genetic material with other organisms, makes many more individuals. Because in this type of reproduction, no exchanges of genetic material occur, the genes of offspring are identical to the parent only. Lack of genetic variation in the organism having asexual reproduction is a disadvantage because it reduces the chance of adaptation. So withstanding in environmental changes is less. However, as opposed to sexual reproduction, the asexual reproducing organisms are grown faster and they produce large quantities of offspring. Rapid reproduction increases the chance of mutation error in genetic material. Very few bacteria can survive mutations; however, those individuals remain alive rebuild the bacterial population again. Different types of asexual reproduction are binary fission, budding, multiple fission, and spore formation[16,74] (Fig. 4.21).

Binary fission: This is the most common method of cell reproduction in bacteria and archaea groups. In this method of reproduction,

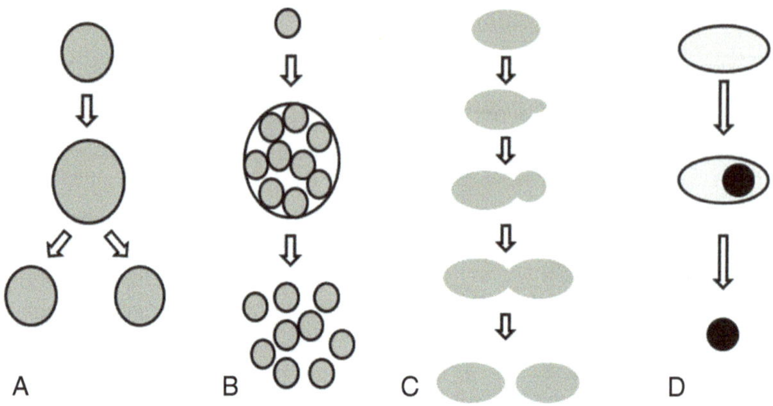

FIG. 4.21 Schematic representation of asexual reproduction in bacteria: **(A)** binary fission, **(B)** multiple fission, **(C)** budding, and **(D)** spore formation.

cell volume increases and then the adult cell divides into two daughter cells.

Budding: In this type of reproduction, initially a small bud projects from mother cell and then the bud enlarges and finally it gets separated from the mother cell.

Multiple fissions: In this process, initially parental cell volume increases and then microbial cell nucleus divides into many nuclei. Later, a thick extracellular matrix (cysts) forms around the cell. Then cytoplasm separates and each cytoplasm gets a single nucleus. Later the cysts break and daughter cells get released.

Spore formation: Spore formation occurs in the harsh environmental condition. Similar to multiple fissions, the parental cell produces many tiny resistant reproductive structures called spore. In appropriate conditions such as the presence of moisture and sufficient nutrient, each spore develops into offspring, which is identical to its parent. Spore formation has been observed in some bacteria, algae, and fungi.

MICROBIAL GROWTH

There are several different processes for bacterial reproduction, but in most common one which is called binary fission, after cell elongation and duplication of its DNA; a bacterial cell divides into two equally sized progeny cells.[2] The time needed for the bacterial division is called generation time. Generation time is important for research studies especially when they aim to utilize or inoculate known numbers of bacterial isolates in an experiment. Generation time can be measured from the bacterial growth curve.

The bacterial growth curve is developed by measuring the rate of increase in cell population over time of incubation. The growth curve actually shows the growth of bacteria when they are cultured in batch culture condition (no fresh medium added and no waste removed during the incubation time). Generally, bacterial growth curve is modeled with four distinct phases (Fig. 4.22):

Lag phase: During this phase, bacteria adjust to a new condition.[3] They synthesize RNA, enzymes, and necessary metabolites needed for their growth in the new environment. Li et al.[75] have shown that the duration of lag phase is proportional to bacterial tolerance against environmental stressor.

Exponential (log) phase: Cells in log phase are the best for performing the experiments because these cells are the healthiest and most uniform one. In this phase, cell division will occur and one bacterial cell becomes two, four, eight, etc. The time needed for the bacterial doubling in log phase can be measured using the following formula[3]:

$$N = N_0 2^n$$

$$g = t/n$$

where

N is the final cell concentration;

N0 is the initial cell concentration;

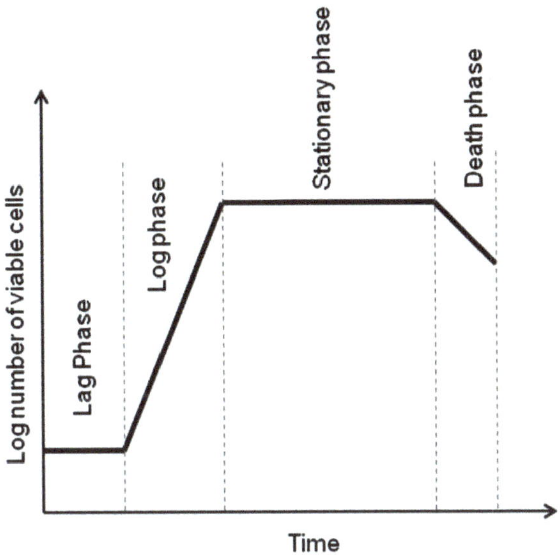

FIG. 4.22 Microbial growth curve and four different growth phases of bacteria in a closed system.

n is the number of generations that occurred between the specified period of time;
g is the generation time; and
t is the specified period of time.

Stationary phase: This phase starts when bacteria that grow in the batch culture use entire nutrient and accumulate all the waste materials. These unfavorable conditions force cells to undergo slower division. Because the number of new cells created is equal to the number of cell death, bacterial growth rate is stabilized in this phase.[3] It is important to note that the bacteria produce their secondary metabolites such as toxins, antibiotics, pigments, and siderophores during the onset of the stationary phase.

Death phase: In this phase, unfavorable condition harms bacteria and individual bacteria completely lose reproduction ability and begin to die.[3] Consider that some bacteria can tolerate unfavorable condition by going through dormancy.[15]

DORMANCY IN BACTERIA

Starvation, oxidative stress, stressful pH, temperature, and salinity are some environmental factors that limit the growth of bacteria. Bacteria can sense these unfavorable conditions and respond to them by going into dormancy, reducing their metabolic activity to an undetectable level in normal laboratory conditions. Dormant cells are silent threats to the industries dealing with the problem of microbial corrosion because these metabolically inactive cells can regrow under appropriate conditions. There are three types of dormant cell:

Endospore: Endospores are structurally, metabolically, and chemically different from vegetative cells.[15] These dense and multilayer bacterial structures are produced by some Gram-positive bacteria such as *Clostridium* and *Bacillus*. Not only persistence of spore but also its stronger attachment to the metal surface is a challenge for MIC treatment.[76]

Persister cell: Persisters are a fraction (up to 1%) of the growing population of microbes in both planktonic and sessile forms. They appear stochastically.[77] They may also form under stressful conditions.[78] These cells are genetically identical to the nonpersister cells but exhibit a drug-tolerant phenotype.[79] They account for antimicrobial tolerance of biofilm. Resuscitation of persister cells occurs in the absence of antibiotics.[80]

Viable but nonculturable cell (VBNC): Similar to other dormant cells, VBNCs are very low in metabolic activity and they cannot divide. Cells in VBNC state are morphologically smaller. VBNC bacteria can remain in this state for over a year. Unlike persisters, VBNC cannot be revived easily. It seems their exposure to some conditions such as rich medium facilitates their resuscitation.[80,81]

PHYSICAL AND CHEMICAL REQUIREMENT FOR BACTERIAL GROWTH

To grow successfully, microbes should adapt to surrounding environments. Microbial growth is greatly affected by the physical and chemical nature of their surroundings. Therefore, understanding of these factors not only helps in the study of the ecological distribution of microorganisms but also in the control of microbial growth.

Microbial cell requires three basic chemical elements (nutrient)[16]:

1. Macronutrients that are needed in large amount for biosynthesis of microbial cell building blocks and also microbial metabolism. Carbon, oxygen, hydrogen, phosphorus, and sulfur are some examples of these essential nutrients.
2. Micronutrients such as iron, manganese, zinc, cobalt, nickel, and copper needed in trace amount for proper function of enzymes and maintenance of protein structure.
3. Growth factors are an essential component of cells that cannot be synthesized by microorganisms. There are three major classes of growth factors: amino acids needed for the synthesis of proteins, purines, and pyrimidines required for nucleic acid synthesis and vitamins, which act as an enzyme cofactor.

The physical factors that affect microbial growth rate are pH, temperature, osmotic pressure, and hydrostatic pressure.[2] Table 4.5 categorizes the bacteria based on their physical requirements.

TABLE 4.5
Categorization of Bacteria Based on Their Physical Requirements

Bacteria	Definition
TEMPERATURE	
Psychrophiles	Grow best at 0°C but optimum growth temperature is 15°C
Psychrotroph	Grow at 0–7°C but optimum growth temperature is between 20 and 30°C
Mesophile	Grow best at moderate temperature around 20–40°C
Thermophile	Grow above optimum 55°C or higher optimum growth temperature is between 55 and 65°C
Hyperthermophile	Optimum growth temperature is between 80 and 110°C
pH	
Acidophiles	Growth optimum between pH 0 and 6
Neutrophiles	Growth optimum between pH 6 and 8
Alkalophiles	Growth optimum between pH 8 and 12
OSMOTIC PRESSURE	
Osmotolerant	Grow at relatively high salt concentration
Halophile	Grow in the presence of salt at concentration between 0.2 and 0.6
OXYGEN CONCENTRATION	
Obligate aerobe	Require oxygen for growth
Microaerophilic	Grow at low concentration of oxygen. High oxygen concentration is toxic for them.
Facultative anaerobe	Grow in the presence and absence of oxygen

BACTERIAL SIGNALING PATHWAYS

Bacteria can survive if they rapidly adapt to changes occurring in their surrounding environment. For this reason, they must be able to access information about their own extracellular and intracellular status. Bacteria use two different signaling pathways, extracellular quorum-sensing signaling and intracellular cyclic dinucleotide signaling, to regulate their rapid reaction.

Quorum-Sensing Signaling

In presence of critical bacterial number that is called a quorum, bacteria start/stop some group activity. But how they sense the quorum? Bacteria produce communicative signals called autoinducers. Autoinducers diffuse across the plasma membrane and get released into the environment. As cell population increases, the concentration of autoinducers increases as well. At high concentration of autoinducers, these molecules diffuse back into individual cells and trigger activation of specific genes[82] (Fig. 4.23). QS system controls many of bacterial activity such as conjugation,

Aerotolerant anaerobe	Grow in the absence of oxygen. Oxygen is not toxic for them
Obligate anaerobe	Grow in the absence of oxygen. Oxygen is toxic for them
Pressure	
Barophilic	Grow in high-pressure environment
Barotolerant	Can grow at high-pressure environment but grow well at atmospheric pressure

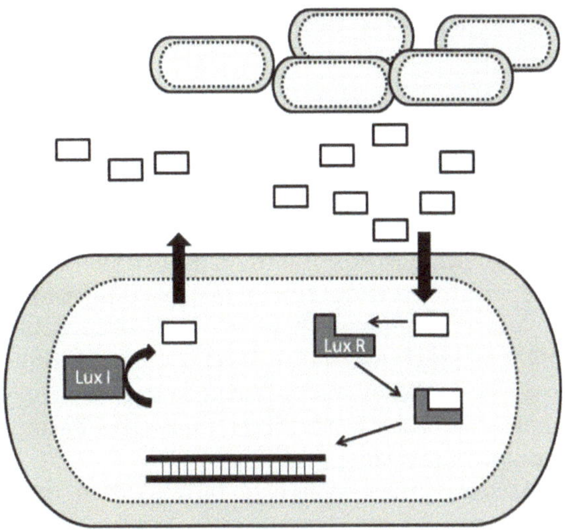

FIG. 4.23 Overview of the quorum sensing mechanism in bioluminescent bacteria. LuxI (a protein that synthesizes the autoinducer) is producing autoinducer molecule, which is diffusing out of the bacterial cell. At certain threshold concentration, autoinducers enter into the cell, bind to LuxR, and activate LuxR (a protein that activates transcription of a set of genes in bioluminescent bacteria).

swarming, sporulation, biofilm formation, production of toxins, and siderophores.[83–88]

Until now, three main QS system has been identified in bacteria: acyl-homoserine lactone (AHL) QS system in Gram-negative bacteria, autoinducing peptide (AIP) QS system in Gram-positive bacteria, and the autoinducer 2 (AI-2) QS system in both Gram-negative and Gram-positive bacteria.[89–91]

Cyclic Dinucleotide Signaling

3,5-Ccyclic diguanylic acid (c-di-GMP) is a unique and widely used secondary messenger (second messengers are intracellular signaling molecules released by the cell in response to exposure to extracellular signaling molecules or the first messengers) in bacteria.[92] Similar to QS molecules, c-di-GMP also controls bacterial activities such as biofilm formation, motility, and bacterial virulence.[93–95]

C-di-GMP is synthesized by diguanylate cyclase enzymes and degraded by phosphodiesterase enzymes. The environmental stimuli such as oxygen, temperature, and nitric oxide have an influence on the activity of these enzymes.[96–99] By affecting enzyme activity, the stimuli alter the concentration of c-di-GMP and mediate the regulation of various proteins in the cell, which subsequently results in bacterial adaptation to the environmental changes.[92]

BACTERIAL LOCOMOTION

Any self-driven motion of bacteria is called bacterial movement. Bacterial movement is vital for their survival. By changing the direction, bacteria move away from repellent and migrate toward attractants. Bacterial movement is playing a role in their virulence and also in their attachment.[100] There are several types of bacteria movement (Fig. 4.24):

Swimming motility: This is a flagellum-dependent movement. This type of locomotion can be seen when bacteria move independently in a liquid medium. Depending on the number of flagella and their arrangements, different run-and-turn patterns can be observed during this type of motility.[30]

Swarming motility: Swarming motility can be seen when bacteria move in association with other cells on the surfaces (or viscous liquid). Once bacteria come into contact with surfaces, their flagellar rotation gets reduced or inhibited. Bacteria can sense this signal and respond to that by inducing swarmer cell differentiation; they elongate, make more flagella or alter flagella's arrangement, secrete wetting compound, and move across the surface in coordinated packs.[31,32]

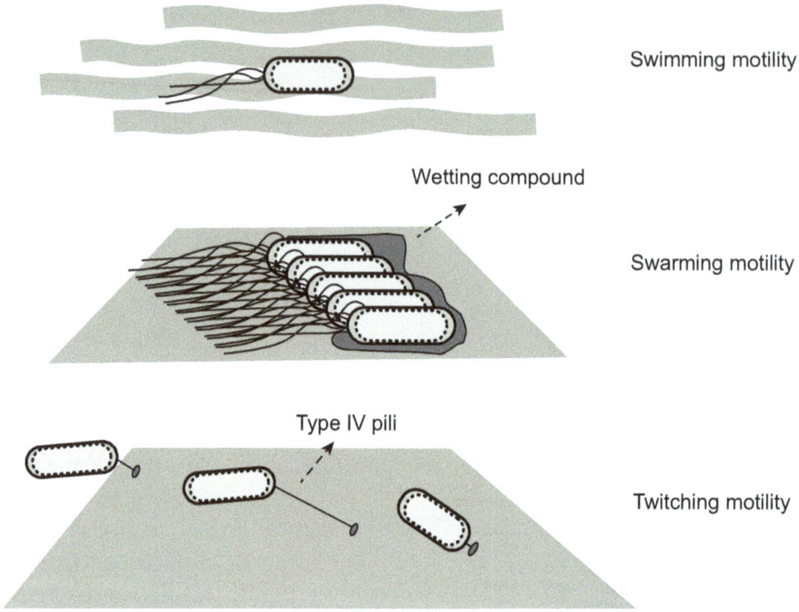

FIG. 4.24 Schematic representation of swimming, swarming, and twitching motility in bacteria.

Swarmer cells are able to reversibly attach to the surface and initiate the formation of biofilm.

Twitching motility: Unlike swimming and swarming motility, twitching is a flagella-independent form of bacterial movement. In this type of movement, bacteria translocate on the solid or semisolid surfaces with the help of type IV pili.[101] During this type of movement, type IV pili act as a grappling hook; it extends, binds to a solid surface, retracts, and pulls bacteria forward. Twitching motility not only is needed for bacterial movement, but also it is necessary for bacterial colonization.[101]

Gliding motility: Gliding is a smooth translocation of bacteria on the solid surface having low water content (such as biofilm) without using any bacterial cell appendages such as flagella and pili. Although the exact mechanism of gliding is not well understood, the mechanism of gliding motility is varied from bacteria to bacteria.[102] Gliding motility can mainly be observed in predatory bacteria when they scout for the prey on the surface.[103]

Sliding motility: Sliding motility is passive bacterial movement powered by pushing forces, which are promoted by the production of biosurfactant, extracellular polymeric substances, hydrophobic protein, and glycopeptidolipids.[104,105]

BIOFILM

In a scientific word, biofilm is a microbial aggregate within an extracellular matrix composed of proteins, exopolysaccharides, and extracellular DNA. Watnick and Kolter[106] analogized biofilm to a city, and they said as we need to take several steps to optimize our lives in a city, bacteria should do the same (Table 4.6).

Bacteria prefer to live in biofilm than being free-floating form because the biofilm matrix protects them from a certain environmental threat such as biocides, antibiotics, and predators. Moreover, biofilm matrix with the ability to entrap and digest nutrients can facilitate microbial nutrient uptake.[107] It also facilitates immobilization of microbes on the material surfaces. Although biofilm is advantageous for bacteria, biofilm formation on any natural or man-made surfaces such as skin, teeth, rocks, cooking vessels, pipes, and boat is harmful. Formation of microbial biofilm on metal has a huge impact on several different industries, as these microbial aggregates may cause corrosion or corrosion inhibition. Therefore, it is necessary to tackle microbial biofilm in industrial systems.

TABLE 4.6
Comparing the Required Steps for Optimization of Life in a City and in a Biofilm

Optimization of Life in a City	Optimization of Life in a Biofilm
Choosing a city to live	**Cell deposition and attachment:** In this step, initially, a free-floating bacterium collides with a preconditioned surface and its motility gets slow. The bacterium forms a transient association with the surface and/or other microbes previously attached to the same surface. This transient association allows the bacterium to search for a place to settle down.
Selecting the neighborhood in the city that best suits our needs	**Colonization:** The attached bacterium produces exopolysaccharide and forms a stable association with the surface and also with other surface-bound microorganisms.
Making our home among the homes of many others	**Maturation:** Cell density increases, EPS fully encloses all the bacterial cells, and biofilm forms a complex, dynamic community.
Leaving the city when life is sour	**Detachment and sloughing:** Occasionally, individual cells or pieces of biofilm detach from the biofilm matrix and they get released to the environment as a form of active dispersal.

INHIBITION OF MICROBES

Significant advances in microbiology demonstrated that the control of microbial growth is necessary. For instance, many studies performed on microbiologically influenced corrosion occurred in various different industries showed that the inhibition of microbial growth could prevent undesirable consequences of the presence of microorganisms or their product on the metals. Generally, controlling microbial growth can be obtained in two ways:

1. By killing microorganisms
2. By inhibiting the growth of microorganisms

Based on the mentioned functioning, agents used for control of microbes are divided into two groups:

1. -cidal agents: The agents that kill bacteria are referred to as cidal agents. For example, bacteriocidal are agents that eliminate bacteria. Fungicides are those agents that kill fungi.
2. -static agents: The agents that just inhibit the growth of microorganisms are called as static agents. Bacteriostatic are agents that inhibit the growth of bacterial cell without killing them.

Selection of microbial control method depends on the situation. Types of microbes present in the environment, the extent of contamination, environmental conditions, and potential risk a control method may have for ecology are some of the factors, which should be considered during the selection of control method. Microbes can be controlled by both physical and chemical means. However, the most common method used in industries for control of microbial growth is the chemical one (use of biocide). While applying biocide, consider the effect of following factors as they have important impacts on treatment efficiency.

1. Concentration of an antimicrobial agent: To obtain the best result, the optimal concentration of antimicrobial agent is needed. Although an increase in concentration or intensity of antimicrobial agents improves the activity of the killing agents, it is not always the case. Sometimes an agent more effectively removes the organism at a lower concentration.

2. Population size: Inhibition of larger population of bacteria needs a higher concentration of antimicrobial agents and a longer period of time than the smaller one.
3. Population composition: Bacterial sensitivity to antimicrobial agent varies remarkably. For instance, Gram-negative bacteria are more susceptible to antimicrobial than the Gram-positive one. Vegetative cells are more sensitive to antimicrobial compound than their dormant compartment. Antimicrobial agents are more effective on planktonic bacteria than their biofilm. Sensitivity of aerobic bacteria to antimicrobial agents is higher than the anaerobic one.
4. Duration of exposure: Longer exposure of microbes to antimicrobial agents increases the efficiency of the antimicrobial agents.
5. Temperature: Increase in temperature improves the antimicrobial activity of the biocidal agents.
6. pH: The activity of many of antimicrobial substances is pH dependent. So maximum activity of an agent can be seen at optimum pH. For instance, glutaraldehyde and cationic biocides (chlorhexidine, QACs) are most active at alkaline pH, whereas hypochlorites and phenolics are most potent at acid pH.
7. Presence of interfering substance: Presence of some organic substance may reduce the effectiveness of antimicrobial agents. For instance, chlorine is inactivated in presence of organic matter. In addition, organic materials protect bacteria from toxic chemicals. For instance, in biofilm, EPS protects the underlying bacteria.

Although considering the above factors can enhance the activity of biocides, biocidal agents are not so effective in biofilm removal. One reason for failure in biocidal removal of biofilm is due to low penetrate of chemical agents into biofilm. The lower growth rate of bacteria and physiologic changes occurring in bacteria that are living in the biofilm are other factors affecting successful biocidal treatment of biofilm.[108] Although, because of these obstacles, biofilm removal is a challenging issue for industries, note that by choosing appropriate chemical agents, you can greatly improve the treatment efficacy. Application of adjuvant along with biocides can enhance the result of biofilm treatment. Biofilm-dispersing agents that improve the accessibility of biocide to bacteria and alter the physiological status of bacteria are other chemical agents, which can be used for effective result.

MICROBIAL INTERACTION

Before talking about different types of microbial interaction, we first explain few terms that are used regularly in microbial ecology.

Microbial ecology: The interactions of microorganisms with each other and with their nonliving physical environment.

Microbial symbioses: Any type of microbial interactions with other living organisms.

Symbiont: Any organisms living in symbiosis.

Microbial population: Group of similar organisms.

Microbial communities: Mixtures of different microbial populations.

In a natural environment, microbes continuously interact with other organisms and their environment. These interactions have an impact on the microbial population and consequently on the functioning of ecosystems. Therefore, knowledge of microbial interactions is import, as it facilitates the understanding of microbial contribution in ecosystems. This information can be used for regulating microbial behavior. The interaction between two microbial populations can positively or negatively affect one or both populations (Fig. 4.25). Some example of microbial interactions is given below:

Mutualism

It is an obligatory interaction between two organisms from different populations in which an individual symbiont benefits from the activity of other. One type of mutualism interaction is syntrophism in which growth of one species depends on the product of other organisms.[16] Santana et al.[109] showed that the *Stenotrophomonas maltophilia* or *Microbacterium* has mutualistic interaction with *Desulfovibrio vulgaris* Hildenborough.

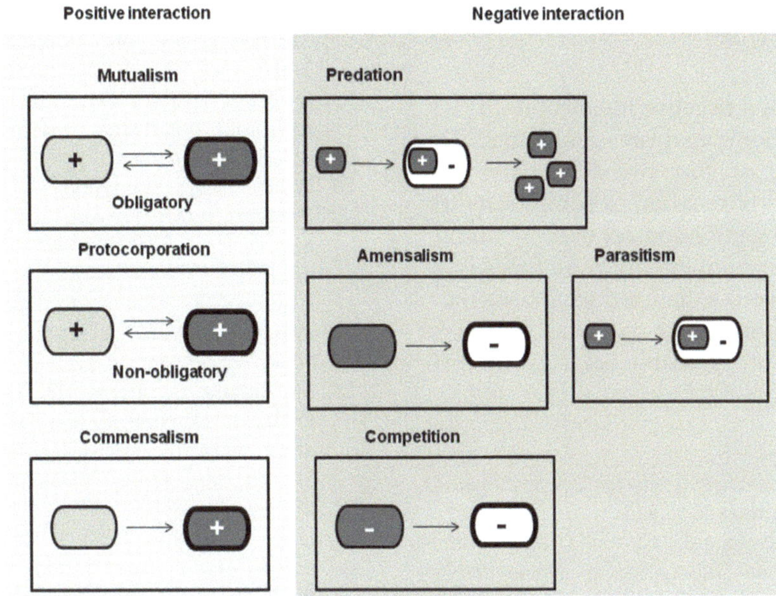

FIG. 4.25 Seven distinct types of positive (mutualism, protocorporation, and commensalism) and negative (predation, amensalism, parasitism, and competition) social interactions in a bacterial population.

These two bacteria produce pyruvate and acetate (carbon sources for *D. vulgaris*) from the carbohydrate sources, which cannot be utilized by *D. vulgaris*.

Protocorporation

Similar to mutualism, in protocorporation, two organisms from different populations benefit from the activity of each other. However, as opposed to mutualisms, this interaction is not obligatory.[16] Therefore, when two organisms get separated if the resources are provided in the environment, then each symbiont can grow independently. An example of protocorporation occurs between sulfate-reducing bacteria and sulfur-oxidizing bacteria.

Commensalism

It is a positive interaction between two microorganisms in which one benefits (commensal) and the other one (host) neither benefits nor gets harmed. Commensal and host can live independently.[16]

Predation

Predation is a negative interaction between two organisms in which one microorganism that is called predator kills and eats the other one called prey.[16] An interesting example of predation in the bacterial community is predation of bacteria by lytic phage or *Bdellovibrio* and like organisms.

Parasitism

In this type of relationship, one organism called parasite lives on/in another organism called host. The parasite may or may not kill the host. In this type of interaction, only one symbiont that is parasite benefits and the host is usually harmed.[16] Parasitic interaction can be observed when lysogenic phage lives in a bacterial host. Although in many cases lysogenic phages harm the host, there are few examples showing that in this type of interactions bacteria benefit from the parasites. For instance, phage infection of bacteria *Corynebacterium diphtheria* induces the toxin production by bacteria.[110]

Amensalism

Amensalism is another negative interaction between two organisms. In amensalism, one organism inhibits or destroys the other one by the production of specific compounds such as toxin, antibiotics, metabolic product, etc.[16] Amensalism can be seen when *Streptomyces lunalinharesii* strain 235 produces antimicrobial compounds against sulfate-reducing bacteria.[111]

Competition

Competition is a negative interaction. In competition, both organisms get harmed. This interaction occurs when two different organisms in a population or community try to acquire the same resources used by both.[16] Very important competition in oil fields is occurring between sulfate-reducing bacteria and heterotrophic nitrate-reducing bacteria for carbon sources (degradable organics carbon).

REFERENCES

1. Ivanov V. *Environmental Microbiology for Engineers*. 2nd ed. CRC press; 2015:413.
2. Pandey A, Teixeira JAC. *Current Developments in Biotechnology and Bioengineering: Foundations of Biotechnology and Bioengineering*. Elsevier; 2016: 298.
3. Hogg S. *Essential Microbiology*. John Wiley & Sons; 2013.
4. A. V. Finkelstein and O. B. B. Ptitsyn, Protein Physics, p. 530.
5. Brändén CI, Tooze J. *Introduction to Protein Structure*. Garland Pub; 1999:410.
6. Thomas WE, Trintchina E, Forero M, Vogel V, Sokurenko EV. Bacterial adhesion to target cells enhanced by shear force. *Cell*. 2002;109(7): 913–923.
7. Alberts B, Johnson A, Lewis J, Raff M, Roberts K, Walter P. *The Lipid Bilayer*. 2002.
8. Jamal M, Ahmad W, Andleeb S, et al. Bacterial biofilm and associated infections. *J Chin Med Assoc*. 2018;81(1):7–11.
9. Limoli DH, Jones CJ, Wozniak DJ. Bacterial extracellular polysaccharides in biofilm formation and function. *Microbiol Spectr*. 2015;3(3).
10. Alberts B, Johnson A, Lewis J, Raff M, Roberts K, Walter P. *The Structure and Function of DNA*. 2002.
11. Ghaz-Jahanian MA, Khodaparastan F, Berenjian A, Jafarizadeh-Malmiri H. Influence of small RNAs on biofilm formation process in bacteria. *Mol Biotechnol*. 2013;55(3):288–297.
12. Lodish H, Berk A, Zipursky SL, Matsudaira P, Baltimore D, Darnell J. *The Three Roles of RNA in Protein Synthesis*. New York: W. H. Freeman; 2000.
13. Chambers JR, Sauer K. Small RNAs and their role in biofilm formation. *Trends Microbiol*. 2013; 21(1):39–49.
14. Sattley WM, Madigan MT. *"Microbiology," in ELS*. Chichester, UK: John Wiley & Sons, Ltd; 2015: 1–10.
15. Dworkin J, Shah IM. Exit from dormancy in microbial organisms. *Nat Rev Microbiol*. 2010; 8(12):890–896.
16. Willey JM, Sherwood L, Woolverton CJ, Prescott LM. *Prescott, Harley, and Klein's Microbiology*. McGraw-Hill Higher Education; 2008.
17. Walker GM, White NA. *"Introduction to Fungal Physiology," in Fungi*. Hoboken, NJ, USA: John Wiley & Sons, Inc; 2017:1–35.
18. McNamara CJ, Perry TD, Leard R, Bearce K, Dante J, Mitchell R. Corrosion of aluminum alloy 2024 by microorganisms isolated from aircraft fuel tanks. *Biofouling*. 2005;21(5–6): 257–265.
19. Little B, Staehle R, Davis R. Fungal influenced corrosion of post-tensioned cables. *Int Biodeterior Biodegrad*. 2001;47(2):71–77.
20. Pintado J, Montero F. Underground biodeterioration of medium tension electric cables. *Int Biodeterior Biodegrad*. 1992;29(3–4):357–365.
21. Larkum AW. Photosynthesis and light harvesting in algae. In: *The Physiology of Microalgae*. Cham: Springer International Publishing; 2016:67–87.
22. Mieszkin S, Callow ME, Callow JA. Interactions between microbial biofilms and marine fouling algae: a mini review. *Biofouling*. 2013;29(9): 1097–1113.
23. Pauli W, Jax K, Berger S. Protozoa in wastewater treatment: function and importance. In: *Biodegradation and Persistance*. Berlin/Heidelberg: Springer-Verlag; 2001:203–252.
24. Beaudoin DJ, Carmichael CA, Nelson RK, Reddy CM, Teske AP, Edgcomb VP. Impact of protists on a hydrocarbon-degrading bacterial community from deep-sea Gulf of Mexico sediments: a microcosm study. *Deep Sea Res Part II: Top Stud Oceanogr*. 2016;129:350–359.
25. Du Toit A. The language of phages. *Nat Rev Microbiol*. 2017;15(3):135.
26. Howard-Varona C, Hargreaves KR, Abedon ST, Sullivan MB. Lysogeny in nature: mechanisms, impact and ecology of temperate phages. *ISME J*. 2017;11(7):1511–1520.
27. Łoś M, Wegrzyn G. Pseudolysogeny. *Adv Virus Res*. 2012;82:339–349.
28. Kojima S, Blair DF. The bacterial flagellar motor: structure and function of a complex molecular machine. *Int Rev Cytol*. 2004;233:93–134.

29. Berg HC, Anderson RA. Bacteria swim by rotating their flagellar filaments. *Nature*. 1973;245(5425): 380–382.
30. Hintsche M, Waljor V, Großmann R, et al. A polar bundle of flagella can drive bacterial swimming by pushing, pulling, or coiling around the cell body. *Sci Rep*. 2017;7(1):16771.
31. Belas R. Biofilms, flagella, and mechanosensing of surfaces by bacteria. *Trends Microbiol*. 2014;22(9): 517–527.
32. Song F, Brasch ME, Wang H, Henderson JH, Sauer K, Ren D. How bacteria respond to material stiffness during attachment: a role of *Escherichia coli* flagellar motility. *ACS Appl Mater Interfaces*. 2017;9(27):22176–22184.
33. Ottow JCG. Ecology, physiology, and genetics of fimbriae and pili. *Annu Rev Microbiol*. 1975; 29(1):79–108.
34. Utada AS, Bennett RR, Fong JCN, et al. *Vibrio cholerae* use pili and flagella synergistically to effect motility switching and conditional surface attachment. *Nat Commun*. 2014;5(1):4913.
35. O'Toole GA, Kolter R. Flagellar and twitching motility are necessary for *Pseudomonas aeruginosa* biofilm development. *Mol Microbiol*. 1998;30(2): 295–304.
36. Sheetz MP, Merz AJ, So M. Pilus retraction powers bacterial twitching motility. *Nature*. 2000; 407(6800):98–102.
37. Zogaj X, Nimtz M, Rohde M, Bokranz W, Romling U. The multicellular morphotypes of *Salmonella typhimurium* and *Escherichia coli* produce cellulose as the second component of the extracellular matrix. *Mol Microbiol*. 2001;39(6): 1452–1463.
38. Reguera G, McCarthy KD, Mehta T, Nicoll JS, Tuominen MT, Lovley DR. Extracellular electron transfer via microbial nanowires. *Nature*. 2005; 435(7045):1098–1101.
39. Sherar B, Power I, Keech P, Mitlin S, Southam G, Shoesmith D. Characterizing the effect of carbon steel exposure in sulfide containing solutions to microbially induced corrosion. *Corrosion Sci*. 2011;53(3):955–960.
40. Silhavy TJ, Kahne D, Walker S. The bacterial cell envelope. *Cold Spring Harb Perspect Biol*. 2010; 2(5):a000414.
41. Vollmer W, Blanot D, De Pedro MA. Peptidoglycan structure and architecture. *Fems Microbiol Rev*. 2008;32(2):149–167.
42. Neuhaus FC, Baddiley J. A continuum of anionic charge: structures and functions of D-alanyl-teichoic acids in gram-positive bacteria. *Microbiol Mol Biol Rev*. 2003;67(4):686–723.
43. Swoboda JG, Campbell J, Meredith TC, Walker S. Wall teichoic acid function, biosynthesis, and inhibition. *Chembiochem*. 2010;11(1):35–45.
44. Frirdich E, Whitfield C. Review: lipopolysaccharide inner core oligosaccharide structure and outer membrane stability in human pathogens belonging to the Enterobacteriaceae. *J Endotoxin Res*. 2005;11(3):133–144.
45. Shang D, Zhang Q, Dong W, Liang H, Bi X. The effects of LPS on the activity of Trp-containing antimicrobial peptides against gram-negative bacteria and endotoxin neutralization. *Acta Biomaterialia*. 2016;33:153–165.
46. Miller SI, Salama NR. The gram-negative bacterial periplasm: size matters. *PLoS Biol*. 2018;16(1): e2004935.
47. Yang DC, Blair KM, Salama NR. Staying in shape: the impact of cell shape on bacterial survival in diverse environments. *Microbiol Mol Biol Rev*. 2016;80(1):187–203.
48. Blumberg PM, Strominger JL. Interaction of penicillin with the bacterial cell: penicillin-binding proteins and penicillin-sensitive enzymes. *Bacteriol Rev*. 1974;38(3):291–335.
49. Gross M, Cramton SE, Götz F, Peschel A. Key role of teichoic acid net charge in *Staphylococcus aureus* colonization of artificial surfaces. *Infect Immun*. 2001;69(5):3423–3426.
50. van Loosdrecht MC, Lyklema J, Norde W, Schraa G, Zehnder AJ. The role of bacterial cell wall hydrophobicity in adhesion. *Appl Environ Microbiol*. 1987;53(8):1893–1897.
51. Schultze-Lam S, Urrutia-Mera M, Beveridge TJ. Metal and silicate sorption and subsequent mineral formation on bacterial surfaces: subsurface implications. In: *Metal Contaminated Aquatic Sediments*. Routledge; 2018:111–147.
52. Campos MA, Vargas MA, Regueiro V, Llompart CM, Albertí S, Bengoechea JA. Capsule polysaccharide mediates bacterial resistance to antimicrobial peptides. *Infect Immun*. 2004; 72(12):7107–7114.
53. Wang H, Wilksch JJ, Strugnell RA, Gee ML. Role of capsular polysaccharides in biofilm formation: an AFM nanomechanics study. *ACS Appl Mater Interfaces*. 2015;7(23):13007–13013.
54. Beveridge TJ, Graham LL. Surface layers of bacteria. *Microbiol Rev*. 1991;55(4):684–705.
55. Sleytr UB, Beveridge TJ. Bacterial S-layers. *Trends Microbiol*. 1999;7(6):253–260.

56. Das T, Sehar S, Manefield M. The roles of extracellular DNA in the structural integrity of extracellular polymeric substance and bacterial biofilm development. *Environ Microbiol Rep*. 2013;5(6): 778–786.
57. Dominiak DM, Nielsen JL, Nielsen PH. Extracellular DNA is abundant and important for microcolony strength in mixed microbial biofilms. *Environ Microbiol*. 2011;13(3):710–721.
58. Harmsen M, Lappann M, Knøchel S, Molin S. Role of extracellular DNA during biofilm formation by Listeria monocytogenes. *Appl Environ Microbiol*. 2010;76(7):2271–2279.
59. Hayes F. The function and organization of plasmids. In: *E. coli Plasmid Vectors*. New Jersey: Humana Press; 2003:1–18.
60. Narenkumar J, Madhavan J, Nicoletti M, Benelli G, Murugan K, Rajasekar A. Role of bacterial plasmid on biofilm formation and its influence on corrosion of engineering materials. *J of Bio- and Tribo-corrosion*. 2016;2(4):24.
61. Lindsay JA. *Staphylococcus aureus* genomics and the impact of horizontal gene transfer. *Int J Med Microbiol*. 2014;304(2):103–109.
62. Hannan S, Ready D, Jasni AS, Rogers M, Pratten J, Roberts AP. Transfer of antibiotic resistance by transformation with eDNA within oral biofilms. *FEMS Immunol Med Microbiol*. 2010;59(3): 345–349.
63. Donlan RM. Biofilms: microbial life on surfaces. *Emerg Infect Dis*. 2002;8(9):881–890.
64. Hausner M, Wuertz S. High rates of conjugation in bacterial biofilms as determined by quantitative in situ analysis. *Appl Environ Microbiol*. 1999; 65(8):3710–3713.
65. Ghigo J-M. Natural conjugative plasmids induce bacterial biofilm development. *Nature*. 2001; 412(6845):442–445.
66. Khosravi-Darani K, Mokhtari Z-B, Amai T, Tanaka K. Microbial production of poly(hydroxybutyrate) from C1 carbon sources. *Appl Microbiol Biotechnol*. 2013;97(4):1407–1424.
67. Friedrich CG, Bardischewsky F, Rother D, Quentmeier A, Fischer J. Prokaryotic sulfur oxidation. *Curr Opin Microbiol*. 2005;8(3): 253–259.
68. Kowalchuk GA, Stephen JR. Ammonia-oxidizing bacteria: a model for molecular microbial ecology. *Annu Rev Microbiol*. 2001;55(1):485–529.
69. Reed BC, DiChristina TJ, Cooper RE, Goff JL, Sekar R. Breathing iron: molecular mechanism of microbial iron reduction by Shewanella oneidensis. In: *Manual of Environmental Microbiology*. 4th ed. American Society of Microbiology; 2016:5.2.1-1–5.2.1-13.
70. Moreno-Vivián C, Cabello P, Martínez-Luque M, Blasco R, Castillo F. Prokaryotic nitrate reduction: molecular properties and functional distinction among bacterial nitrate reductases. *J Bacteriol*. 1999;181(21):6573–6584.
71. Rabus R, Hansen TA, Widdel F. Dissimilatory sulfate- and sulfur-reducing prokaryotes. In: *The Prokaryotes*. New York, NY: Springer New York; 2006:659–768.
72. Figueroa RI, Bravo I, Garcés E. The significance of sexual versus asexual cyst formation in the life cycle of the noxious dinoflagellate *Alexandrium peruvianum*. *Harmful Algae*. 2008;7(5): 653–663.
73. O'Gorman CM, Fuller HT, Dyer PS. Discovery of a sexual cycle in the opportunistic fungal pathogen *Aspergillus fumigatus*. *Nature*. 2009;457(7228): 471–474.
74. Angert ER. Alternatives to binary fission in bacteria. *Nat Rev Microbiol*. 2005;3(3):214–224.
75. Li B, Qiu Y, Shi H, Yin H. The importance of lag time extension in determining bacterial resistance to antibiotics. *Analyst*. 2016;141(10):3059–3067.
76. Harimawan A, Zhong S, Lim C-T, Ting Y-P. "Adhesion of *Bacillus subtilis* spores and vegetative cells onto stainless steel – DLVO theories and AFM spectroscopy. *J Colloid Interface Sci*. 2013;405: 233–241.
77. Germain E, Roghanian M, Gerdes K, Maisonneuve E. Stochastic induction of persister cells by HipA through (p)ppGpp-mediated activation of mRNA endonucleases. *Proc Natl Acad Sci U S A*. 2015;112(16):5171–5176.
78. Bernier SP, Lebeaux D, DeFrancesco AS, et al. Starvation, together with the SOS response, mediates high biofilm-specific tolerance to the fluoroquinolone ofloxacin. *PLoS Genet*. 2013;9(1): e1003144.
79. Wood TK, Knabel SJ, Kwan BW. Bacterial persister cell formation and dormancy. *Appl Environ Microbiol*. 2013;79(23):7116–7121.
80. Ayrapetyan M, Williams TC, Oliver JD. Bridging the gap between viable but non-culturable and antibiotic persistent bacteria. *Trends Microbiol*. 2015;23(1):7–13.
81. Ramamurthy T, Ghosh A, Pazhani GP, Shinoda S. Current perspectives on viable but non-culturable (VBNC) pathogenic bacteria. *Front Public Health*. 2014;2:103.

82. Williams P. Quorum sensing, communication and cross-kingdom signalling in the bacterial world. *Microbiology*. 2007;153(12):3923–3938.
83. Rösch TC, Graumann PL. Induction of plasmid conjugation in *Bacillus subtilis* is bistable and driven by a direct interaction of a Rap/phr quorum-sensing system with a master Repressor. *J Biol Chem*. 2015;290(33):20221–20232.
84. Nickzad A, Lépine F, Déziel E. Quorum sensing controls swarming motility of *Burkholderia glumae* through regulation of Rhamnolipids. *PLoS one*. 2015;10(6):e0128509.
85. Perchat S, Talagas A, Poncet S, et al. How quorum sensing connects sporulation to necrotrophism in *Bacillus thuringiensis*. *Plos Pathog*. 2016;12(8): e1005779.
86. Tseng BS, Majerczyk CD, Passos da Silva D, Chandler JR, Greenberg EP, Parsek MR. Quorum sensing influences *Burkholderia thailandensis* biofilm development and matrix production. *J Bacteriol*. 2016;198(19):2643–2650.
87. Essigmann HT, Darkoh C, McHugh EE, Brown EL. The *Clostridium difficile* quorum-sensing molecule alters the *Staphylococcus aureus* toxin expression profile. *Int J Antimicrob Agents*. 2017;49(3): 391–393.
88. Popat R, Harrison F, da Silva AC, et al. Environmental modification via a quorum sensing molecule influences the social landscape of siderophore production. *Proc Biol Sci*. 2017;284(1852): 20170200.
89. McNab R, Ford SK, El-Sabaeny A, Barbieri B, Cook GS, Lamont RJ. LuxS-based signaling in *Streptococcus gordonii*: autoinducer 2 controls carbohydrate metabolism and biofilm formation with *Porphyromonas gingivalis*. *J Bacteriol*. 2003;185(1): 274–284.
90. Miller ST, Xavier KB, Campagna SR, et al. *Salmonella typhimurium* Recognizes a chemically distinct form of the bacterial quorum-sensing signal AI-2. *Mol Cell*. 2004;15(5):677–687.
91. Miller MB, Bassler BL. Quorum sensing in bacteria. *Annu Rev Microbiol*. 2001;55(1):165–199.
92. Camilli A, Bassler BL. Bacterial small-molecule signaling pathways. *Science*. 2006;311(5764): 1113–1116.
93. Ha D-G, O'Toole GA. c-di-GMP and its effects on biofilm formation and dispersion: a *Pseudomonas aeruginosa* review. *Microbiol Spectr*. 2015;3(2). MB-0003-2014.
94. Kuchma SL, Delalez NJ, Filkins LM, Snavely EA, Armitage JP, O'Toole GA. Cyclic di-GMP-mediated repression of swarming motility by *Pseudomonas aeruginosa* PA14 requires the MotAB stator. *J Bacteriol*. 2015;197(3):420–430.
95. Ryan RP. Cyclic di-GMP signalling and the regulation of bacterial virulence. *Microbiology*. 2013; 159(Pt 7):1286–1297.
96. Tuckerman JR, Gonzalez G, Sousa EHS, et al. An oxygen-sensing diguanylate cyclase and phosphodiesterase couple for c-di-GMP control. *Biochemistry*. 2009;48(41):9764–9774.
97. Xiao G, Kong L, Che R, et al. Identification and characterization of c-di-GMP metabolic enzymes of *Leptospira interrogans* and c-di-GMP Fluctuations after thermal shift and infection. *Front Microbiol*. 2018;9:764.
98. Cutruzzolà F, Frankenberg-Dinkel N. Origin and impact of nitric oxide in *Pseudomonas aeruginosa* biofilms. *J Bacteriol*. 2016;198(1):55–65.
99. Koestler BJ, Waters CM. Exploring environmental control of cyclic di-GMP signaling in *Vibrio cholerae* by using the ex vivo lysate cyclic di-GMP assay (TELCA). *Appl Environ Microbiol*. 2013;79(17): 5233–5241.
100. Josenhans C, Suerbaum S. The role of motility as a virulence factor in bacteria. *Int J Med Microbiol*. 2002;291(8):605–614.
101. Mattick JS. Type IV pili and twitching motility. *Annu Rev Microbiol*. 2002;56(1):289–314.
102. Nan B, Zusman DR. Novel mechanisms power bacterial gliding motility. *Mol Microbiol*. 2016; 101(2):186–193.
103. Lambert C, Fenton AK, Hobley L, Sockett RE. Predatory *Bdellovibrio bacteria* use gliding motility to scout for prey on surfaces. *J Bacteriol*. 2011; 193(12):3139–3141.
104. Hölscher T, Kovács ÁT. Sliding on the surface: bacterial spreading without an active motor. *Environ Microbiol*. 2017;19(7):2537–2545.
105. Kinsinger RF, Shirk MC, Fall R. Rapid surface motility in Bacillus subtilis is dependent on extracellular surfactin and potassium ion. *J Bacteriol*. 2003;185(18):5627–5631.
106. Watnick P, Kolter R. Biofilm, city of microbes. *J Bacteriol*. 2000;182(10):2675–2679.
107. Flemming H-C, Wingender J, Szewzyk U, Steinberg P, Rice SA, Kjelleberg S. Biofilms: an emergent form of bacterial life. *Nat Rev Microbiol*. 2016;14(9):563–575.

108. Davies D. Understanding biofilm resistance to antibacterial agents. *Nat Rev Drug Discov*. 2003; 2(2):114–122.
109. Santana MM, Portillo MC, Gonzalez JM. Mutualistic growth of the sulfate-reducer *Desulfovibrio vulgaris* Hildenborough with different carbohydrates. *Microbiology*. 2012;81(6):663–668.
110. Zajdowicz SLW, Holmes RK. *Phage Conversion and the Role of Bacteriophage and Host Functions in Regulation of Diphtheria Toxin Production by Corynebacterium Diphtheriae*. Cham: Springer; 2016:15–45.
111. Rosa J P d, Tibúrcio SRG, Marques JM, Seldin L, Coelho RRR. *Streptomyces lunalinharesii* 235 prevents the formation of a sulfate-reducing bacterial biofilm. *Braz J Microbiol*. 2016;47(3): 603–609.

CHAPTER 5

Biologic Treatment of Corrosion

INTRODUCTION

The idea of exploiting microbes for corrosion protection is not new, and corrosion inhibition using microbes has been under research for several decades. The most important approaches that have been discussed so far are applications of bacteriophages, nitrate-reducing bacteria and regenerative biofilm. Although most of these corrosion inhibition approaches were successful in the laboratory conditions, only very few are being used in industries. Therefore in an attempt to identify the requirement for industrial applicability of these methods, in this chapter, we talk about various potent biologic approaches of corrosion inhibition. We deal with limitations of these approaches and then we give some suggestions to overcome these constraints.

CORROSION INHIBITION BY PREDATORY MICROORGANISMS

Predation is one negative interaction among prokaryotes in which one organism named predator feeds on the other(s) called prey. Because predation always results in the death of a prey, a better understanding of this interaction could direct us to novel strategies for the elimination of microbial biofilm and subsequently for inhibition of microbiologically influenced corrosion.

Bacteria are subjected to various eukaryotic and prokaryotic predators. However, discussion about predatory eukaryotes (predatory protists) is not in the scope of this book, and the central theme in this section is just dedicated to predatory bacteria and bacteriophages.

Corrosion Inhibition by Predatory Bacteria

An introduction about predatory bacteria

Predatory bacteria use four different strategies to invade their prey[1] (Fig. 5.1):

1. Wolfpack or group predation: In this type of predatory approach, predatory cells degrade nearby bacteria by excreting a variety of hydrolytic extracellular enzymes and then they assimilate prey-derived nutrients.
2. Epibiotic predation: In epibiotic predation, predators initially attach to the outer cell wall of a prey and then degrade the host content. This invasion is observed in *Vampirococcus* and *Ensifer*.[2]
3. Direct invasion of cytoplasm: This type of predation has just reported in *Daptobacter*, and there is no more data available on this strategy.
4. Periplasmic predation: In periplasmic invasion, predator sets itself in the prey's periplasmic space. This type of predation is accomplished by *Bdellovibrio* and related organisms are referred to as *Bdellovibrio* and like organisms (BALOs). There are four recognized genera of BALOs: *Bdellovibrio*, *Bacteriolyticum*, *Peredibacter*, and *Halobacteriovorax* (previously named as *Bacteriovorax*)[3].

This strategy is the most extensively studied predatory process both by field observation and by laboratory experiments and therefore will be the focus of this part of the book.

BALOs are ubiquitous in nature. They have been discovered in a wide variety of environments ranging from sweet and salt aquatic environments to terrestrial habitats.[4,5] BALOs are present in both planktonic and sessile forms. They prefer the

Biological Treatment of Microbial Corrosion. https://doi.org/10.1016/B978-0-12-816108-1.00005-7

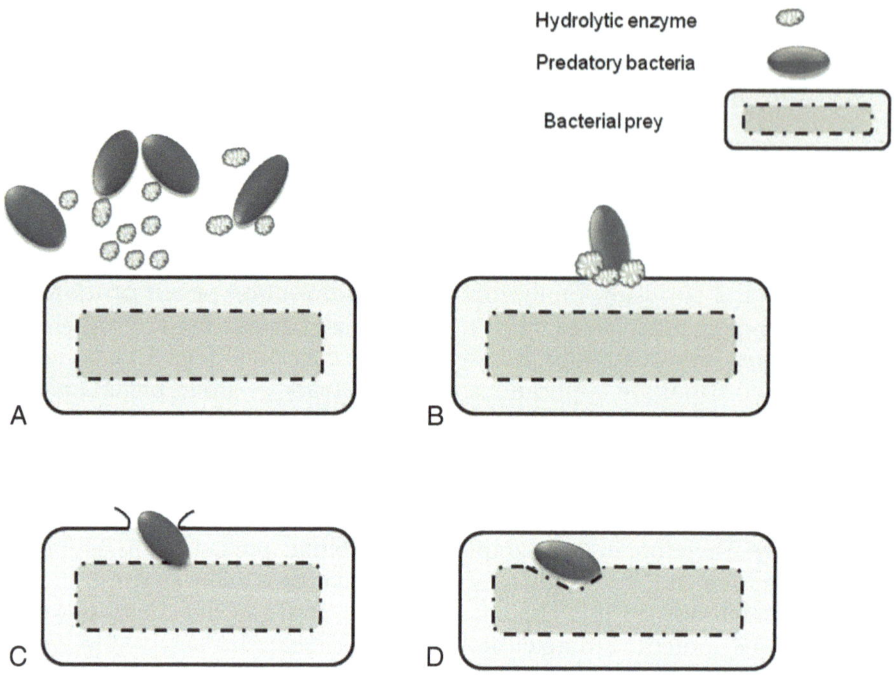

FIG. 5.1 Schematic representation of different predatory strategies used by bacteria: **(A)** wolfpack predation, **(B)** epibiotic predation, **(C)** direct invasion of the cytoplasm, and **(D)** periplasmic predation.

surface association growth compared to free-floating lifestyle because biofilm protects BALOs from unfavorable condition and provides a good source of food (high density of bacteria) for them.[6] *Bdellovibrio bacteriovorus* HD 100 is the most highly studied BALOs. *B. bacteriovorus* HD 100 is a Gram-negative aerobic bacterium with two distinct subpopulations; an obligatory parasitic population is called host-dependent (HD) and a saprophyte is host-independent (HI) derivative.[7] Predatory and HI growth phase of *B. bacteriovorus* is summarized in Fig. 5.2. In HD form, *B. bacteriovorus* uses its single polar flagellum and swims around with high velocity to search for its prey. Upon encountering a suitable prey cell, which apparently occurs by random collision and presence of attractant, *Bdellovibrio* reversibly attaches to its prey. It produces a mixture of hydrolytic enzymes and generates a small opening in the prey cell's outer membrane and peptidoglycan layer. Entry into prey periplasm occurs about 10 min. During this time, *Bdellovibrio* sheds its flagellum. After the entry process, the predator alters the host cell to a spherical structure called bdelloplast. *Bdellovibrio* utilizes prey cytoplasm content and grows until it becomes multiple of its normal size. As the food supplies become exhausted, *Bdellovibrio* dissolves the remaining peptidoglycan and outer membrane layer, which results in the release of progeny cells.[8] This intraperiplasmic phase usually takes about 3–4 h.[9] During this phase, three to six progeny cells will be produced from a single *Escherichia coli* cell.[9] In contrast to host-dependent bacteria, HI subpopulations do not provide their nutrients from the predation. They uptake nutrients that are available in the environment.[7]

Mutational events as well as environmental factors such as nutrient availability, prey abundance, and motility play a role in the variation of the *B. bacteriovorus* population into these two distinct subpopulations.[7,10] In the environment with poor nutrient and high prey availability, *Bdellovibrio* bacteria consume their prey and multiply robustly, whereas in nutrient-rich condition with poor prey population, HD variant of *Bdellovibrio*

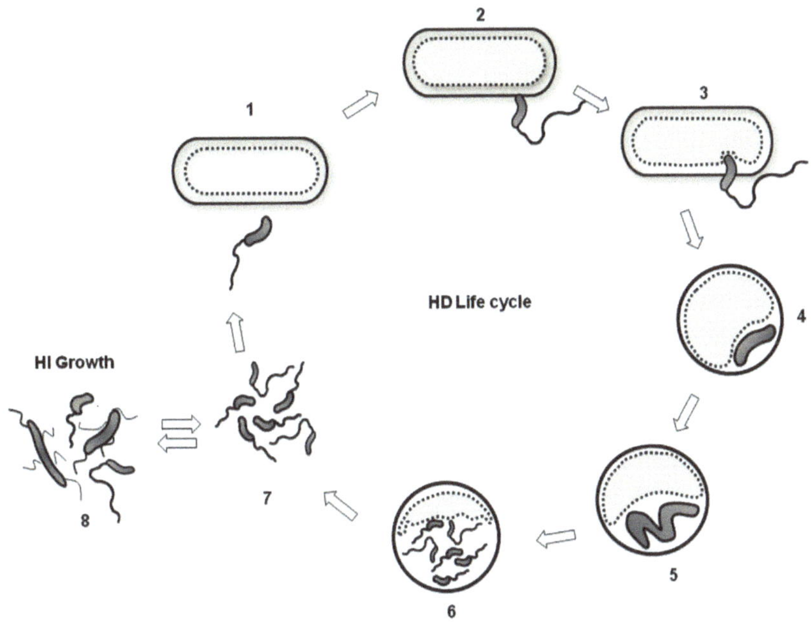

FIG. 5.2 Schematic representation of host-dependent (HD) and host-independent (HI) life cycle of *Bdellovibrio bacteriovorus*. 1. Searching for prey. 2. Attachment to prey. 3. Entry into periplasm. 4. Formation of bdelloplast. 5. Filamentous growth in the periplasm. 6. Segregation of filamentous *Bdellovibrio*. 7. Release of progenies. 8. Short attack-phase-like cells and long filamentous cells in HI phase.

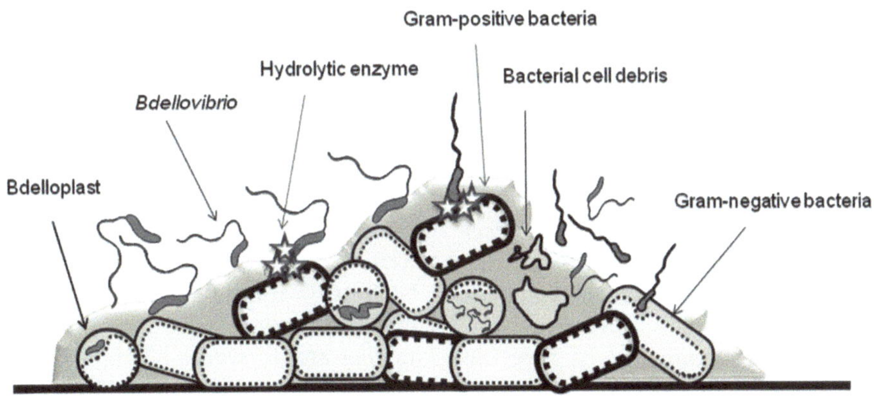

FIG. 5.3 Graphic representation of biofilm removal by both epibiotic and periplasmic predation.

changes into HI form.[7] Switching strategy from HD to HI lifestyle is beneficial for *Bdellovibrio* because it helps the predator to survive when the low number of prey is available.

For many years, scientists believed that BALOs could only attack Gram-negative bacteria; however, recently it has been determined that *B. bacteriovorus* HD100 can effectively attack the Gram-positive bacteria too.[11] *B. bacteriovorus* has dual foraging strategy against Gram-negative and Gram-positive host. This bacterium attacks Gram-positive preys in an epibiotic manner and consumes Gram-negative preys in a periplasmic approach[12] (Fig. 5.3). Host-independent *B. bacteriovorus* HD100 secretes a significant amount of extracellular proteases and DNAse into culture media and degrades the biofilm of Gram-positive *Staphylococcus aureus*. Monnapa et al.[13] showed that the addition of 10% supernatant from HD100 culture could reduce biofilm formation of *S. aureus* by 75% (or greater). This culture supernatant also could disperse the preformed biofilms by 75%. Analysis of supernatant by various techniques revealed the presence of active hydrolytic protease and DNAse enzyme in it.

BALOs are aerobic bacteria. Schoffield et al.[14] showed that BALO strains could survive for up to 9 days under anoxic environments and some BALO strains could grow and attack preys under microaerobic conditions, albeit at a reduced rate compared with fully aerobic tests. The existence of cytochrome oxidase complex, nitrite reductase, and nitric oxide reductase genes identified in the HD100 genome indicates that both microaerophilic and anaerobic respiration can be exploited by the BALOs.[9] These findings expand the potential environments that could support predation by BALOs to microaerophilic and anaerobic habitats.

Predatory bacteria as a corrosion inhibitory agent

BALOs are able to prey upon a diverse group of bacteria in both planktonic and sessile forms.[15] The antibiofilm activity of BALOs highlights the potential application of these bacteria for the treatment of harmful biofilm (Fig. 5.3). In 2016, Qiu and his colleagues investigated the effect of *B. bacteriovorus* on the corrosion of X70 steel exposed to SRB. By inhibiting the SRB growth, which was evident by the decrease in value of optical density and sulfate concentration, *B. bacteriovorus* inhibited corrosion of pipeline steel. In the presence of the predator, corrosion rate declined from 19.17 to 3.75 mg dm^{-2} day^{-1} and corrosion potential shifted in the negative direction.[16]

Using BALOs for corrosion control may be an attractive option for many industries as they can replace biocides. Predatory bacteria are a good option for eradication of thick biofilm because they penetrate deeply into biofilm and eradicate dormant cells.[15,17] They also can invade encapsulated preys.[18] Although most BALOs exhibit a preference for certain species of prey, they attack a wide number of bacteria.[19]

Therefore, these broad-spectrum predators can be used for effective eradication of mix population biofilm. One more advantageous property of BALOs is related to their self-replication ability, which lowers the pulse of injection. BALOs not only can eradicate resistant bacteria but also do not stimulate permanent resistance in preys, a property mentioned as an advantage over phages. Shemesh and Jurkevitch[20] observed that even after inoculation of high concentration of *B. bacteriovorus* SNE into suspension of *Erwinia carotovora* ssp. *carotovora*, a fraction of prey population was alive. They subcultured the remaining prey cells in fresh media for one night and then exposed these prey to *B. bacteriovorus* once again. One hour after second exposure, the prey population decreased by 99.5% and it did not decline later. This result indicates that resistance to predatory bacteria is transient and away from predators, prey, or their descendants regain susceptibility. This transient resistant, which is called plastic resistance, is a phenotypic change and it occurs due to the production of S-layer (and not capsule) on the surface of a host cell.[21] Bacteria can also escape predation through spore formation. Muller et al.[22] investigated the prey-predator interaction between *Myxococcus xanthus* and 15 different species of bacteria (9 strains of *Bacillus subtilis*, 2 strains of *Escherichia coli*, 1 *Salmonella enterica*, 1 *Pseudomonas aeruginosa*, 1 *Staphylococcus aureus*, and 1 *Rhodobacter capsulatus* strain). All the tested bacteria were susceptible to the predation except *B. subtilis* NCIB3610. It has been shown that the prolonged prey-predatory interaction between *Myxococcus xanthus* and *B. subtilis* NCIB3610 induces the formation of megastructure biofilm (very large, a tree-like structure containing many spores in its middle part) by *B. subtilis*.[23] Spore formation allows strain NCIB3610 to resist against the predator. Muller and his team[22] mentioned that entering into dormancy via spore formation is not the only defense mechanism. Production of bacillaene, a secondary metabolite that can inhibit the protein synthesis in prokaryotes, was another strategy by which *B. subtilis* could protect themselves from predation[23] (Fig. 5.4). Persistence of resistant

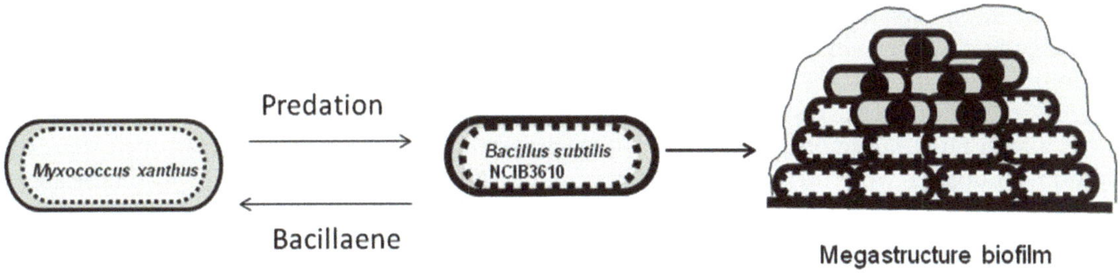

FIG. 5.4 *Bacillus subtilis* NCIB3610 protects itself from *Myxococcus xanthus* predation by two mechanisms: production of bacillaene (for transient resistance) and formation of megastructure biofilm (for long-time resistance).

bacteria could be a concern for MIC treatment as they can reestablish biofilm. However, it has been shown that the presence of a diverse group of prey or predators (and not increase in the incubation timing or reinoculation of a fresh predator) can certainly reduce the formation of plastic resistant.[15,24,25]

Even if BALOs overcome the problem of plastic resistance, they fail to remove biofilm completely because they cannot eradicate the EPS.[15] Incomplete removal of biofilm poses a particular concern if this technique is used for the treatment of MIC. The residual EPS not only facilitates attachment of other bacteria but also provides a patchy and discontinuous film on the metal surface. Patchy film accelerates corrosion process through formation of differential concentration cells. Thus a combination of bacterial predation technique with an effective physical or chemical method to remove EPS is recommended.

Another challenge for the application of BALOs is their host specificity. Although BALOs can infect a wide variety of Gram-negative bacteria, the prey range susceptible to specific BALO strains varies strongly. Van Essche et al.[19] evaluated the susceptibility of six different prey bacteria (*Aggregatibacter actinomycetemcomitans, Eikenella corrodens, Porphyromonas gingivalis, Capnocytophaga sputigena, Prevotella intermedia,* and *Fusobacterium nucleatum*) toward predation by six different BALO strains. Selected predators have shown an individual preference for different prey strains. Although a single BALO strain could attack different species of prey, none of the tested BALO strains covered all the bacterial prey (*B. bacteriovorus* HD100 invaded four of the six tested strains. *Bdellovibrio bacteriovorus* 109J decreased the viability of three of the six tested preys. Each one of the *Bacteriovorus* BEP2 and *Bacteriovorax* FCE was only effective in predation on two of the six strains. *Peredibacter starrii* A3.12 infected only one strain and *Bacteriovorax stolpii* Uki2 did not infect any of the tested pathogens). Host specificity does not preclude predation activity of BALOs; however, it influences both chance and speed of predation. Because the interaction between *Bdellovibrio* and its prey is majorly based on the chance of collision, the probability of a *Bdellovibrio* in finding prey becomes lower as prey density decreases.[26] The problem caused by the host specificity can be solved by using broad-spectrum predators or cocktail of predatory bacteria.

For the successful application of BALOs, it is important to use a suitable concentration of predatory bacteria. Using dead-end microfiltration tests, Kim et al.[27] monitored the effect of *Bdellovibrio bacteriovorus* HD100 on membrane biofouling induced by *Escherichia coli* DH5a in two different statuses of the multiplicity of infection (MOI) within 48 h. The MOI that is defined as the predator to prey cell ratio was 2 (low MOI) and 200 (high MOI). Within the first 24 h, eradication of *E. coli* was superior (around six logs) in high MOI condition as compared with low MOI and control (no predator added) conditions. However, at the end of the experiment (after 48 h), *E. coli* population was nearly identical in both high and low MOI conditions and the number of *E. coli* was four logs lower than the control. Although treatment in both high and low MOI conditions improved membrane performance, examining the effect of individual factors contributing into flux decline (such as concentration polarization, cake formation, and pores blocking) showed a high rate of pore blocking in low MOI condition compared with high MOI condition and control. Higher irreversible fouling attributed to the production of the high amount of *E. coli* debris after predation in low MOI condition.

During the treatment by predatory bacteria, structure of biofilm and its bacterial composition should be considered. Exploitation of BALOs along with predators of Gram-positive bacteria such as *Cupriavidus* and *Myxococcus* is recommended for treatment of biofilm having a high number of Gram-positive bacteria[23,28] (Table 5.1). It is more reasonable to operate predatory bacteria for the elimination of continuous biofilm compared with the patchy and heterogeneous one. When Hol and his coworker[46] examined the effect of spatial structure fragmentation on prey dynamics, they observed that *Bdellovibrio bacteriovorus* HD100 completely eradicates the *E. coli* population in a continuous landscape, whereas a significant part of the prey population in the fragmented landscape persisted over time.

Physical, chemical, and biologic properties of each ecosystem also affect the efficacy of biofilm treatment by BALOs. Presence of bdellophages in natural environment reduces the efficacy of predation as these phages specifically infect *Bdellovibrio bacteriovorus* species.[47] High nutrient availability supports the host-independent growth of *Bdellovibrio* and reduces the treatment efficacy.

TABLE 5.1
List of Different Predatory Bacteria

Predatory Strategy	Predatory Bacteria	Type of Prey	Host Dependency	Ref.
Wolfpack	*Myxococcus xanthus*	G+,G−, Fungi	Facultative	29
Wolfpack, cell contact	*Lysobacter* sp.	G+,G−, *Cyanobacteria*	Facultative	30,31
Wolfpack	*Saprospira albida*	G+,G−, *Cyanobacteria*	—	32
Wolfpack	*Tenacibaculum*	G+,G−	Facultative	33
Wolfpack	*Olleya* sp.	G+,G−	Facultative	33
Wolfpack	*Herpetosiphon* sp.	Unencapsulated bacteria	Facultative	2,34
Epibiotic	*Vampirococcus* sp.	*Chromatium* spp.	Obligate	35
Epibiotic	*Ensifer adhaerens*	G+	Facultative	1
Epibiotic	*Micavibrio* sp.	G−	Obligate	36
Epibiotic	*Cupriavidus necator*	G+,G−	Facultative	28,37
Epibiotic	*Agromyces ramosus*	G+,G−	Facultative	38
Epibiotic	*Streptoverticillium* sp.	G+	Facultative	39
Epibiotic	*Aristabacter necator*	G+,G−, Fungi	Facultative	40
Epibiotic	*Streptomyces* sp.	G+,G−		41
Epibiotic	*Cytophaga* sp.	Cyanobacteria	—	42
Periplasmic, epibiotic	*Bdellovibrio* sp.	G−,G+	Facultative	11
Periplasmic	*Halobacteriovorax* sp.	G−	Obligate	43
Periplasmic	*Bacteriolyticum* sp.	G−	Obligate	44
Direct invasion of cytoplasm	*Daptobacter* sp.	*Chromotium* spp., *Thiocapsa* sp.	Facultative	45

Iebba et al.[12] investigated the predatory behavior of *B. bacteriovorus* against *Pseudomonas aeruginosa* and *Staphylococcus aureus* biofilms in both flow and static conditions. They observed that the predatory activity of *B. bacteriovorus* decreased in flow condition by 38% (*P. aeruginosa*) and 28% (*S. aureus*) compared with "static" condition. Kadouri and Tran[48] reported that the oxygen can stimulate the predation rate. They demonstrated that predators such as *Bdellovibrio bacteriovorus* and *Micavibrio aeruginosavorus* can effectively remove both planktonic and sessile bacteria in aerobic condition, whereas they just attack bacterial biofilm in microaerophilic conditions. Pollution and impurities reduce the chance of collision and impede the predator-prey interactions, a problem that might be decreased by agitation. While treating MIC with combined chemical/biologic (predatory bacteria) techniques, it seems more reasonable to exploit predation technique either before or after chemical treatment because biocides and detergents are having an inhibitory influence on the activity of *Bdellovibrios*.[49]

Corrosion Inhibition by Predatory Phages

The renewed interest in alternative corrosion inhibition methods has turned attention to treatment by bacteriophage. Bacteriophages have the natural capacity to act as antimicrobial agents, and therefore they can be exploited for the elimination of corrosion-causing bacteria. Although ability of phages to infect bacteria in both planktonic and sessile forms makes them ideal candidates for inhibition of MIC, there are some obstacles for phage usage in industries. The resistance of bacteria to phage particle, conversion of lytic phages to lysogenic phages, and narrow host range are some of the most concerning hurdles to address to make substantial impacts in the field.

Narrow host range

As a minimal requirement for phage treatment, a phage should be able to infect its bacterial host. Bacteriophages do not randomly attach to the surface of a host cell; rather, they fasten to specific surface structures called receptor. Receptor structure is so specific that a phage can only attack bacteria having a cell surface that exactly "matches."[50] Eydal et al.[51] isolated lytic phages of *Desulfovibrio aespoeensis* from different boreholes using the most probable number standard method. To find out the host range of phages, they used 6 different species of *Desulfovibrio* and 10 isolated SRB. None of these bacteria were infected by the isolated phage, and the *D. aespoeensis* was the only host.

Theoretically, phage specificity is a disadvantage for treatment of natural biofilm composed of a heterogeneous bacterial population because only a fraction of bacteria in a biofilm can be infected. Nevertheless, in reality it is not an obstacle for the eradication of biofilm. Sillankorva et al.[52] exposed dual-species biofilm of *Pseudomonas fluorescens* and *Staphylococcus lentus* (mostly consisting of *S. lentus*) to phage fIBB-PF7A, which was specific only for *P. fluorescens*. The phage fIBB-PF7A infected its host cell, partially damaged the biofilm structure, and released the nonsusceptible host (*S. lentus*) from biofilm to the planktonic phase. A switch from sessile to planktonic phase may enhance the effectiveness of biocide treatment, as biocides are more powerful on planktonic bacteria.

New techniques that expand the target range can improve the antibiofilm activity of phages. One of these techniques is the use of phages with broader infective spectra. Depending on the specificity of receptors, phages can be categorized into monovalent phages with the capacity to infect a specific bacterial species or strain, and polyvalent phages that are able to infect across different bacterial species or even genera.[53] Polyvalent phages with broad spectra can potentially reduce the number of phages needed for biofilm removal. A study of dual-species biofilm formed by *Pseudomonas fluorescens* (Gram-negative) and *Staphylococcus lentus* (Gram-positive) demonstrated successful disinfection by the action of polyvalent lytic phages.[52] Phages with broad-spectrum activity can be isolated from the reservoir or they can be made through genetic engineering. Marzari et al.[54] showed that grafting the g3p phage protein of one filamentous phage (IKe) to another (fd) enabled an extension of phage

TABLE 5.2
Possible Mechanisms of Improving Biofilm Control by Bacteriophages

Action	Example	Reference
Production of polysaccharide-degrading enzymes	Biofilm of *Escherichia coli* dispersed by genetically modified bacteriophage T7 excreting dispersin B enzyme	55
	K1 capsule in *Escherichia coli* degraded by virion-associated endosialidases enzymes of K1-specific phages	56
Production of proteolytic enzymes	Endolysin enzyme expressed by recombinant bacteriophage phi11- and phi12-hydrolyzed *staphylococcal* biofilms	57
	Virion-associated peptidoglycan hydrolases enzyme from bacteriophage vB_SauS-phiIPLA88 showed lytic activity against *Staphylococcus aureus*	58
Production of anti-QS enzymes	Bacteriophage T7, which was expressing enzyme AiiA lactonase effectively, inhibited the mixed species biofilms of *Pseudomonas aeruginosa* and *Escherichia coli*	59
Delivering antimicrobial agents	M13 phagemid expressing Gef and ChpBK toxin reduced the number of *E. coli* ER2738 in both in vitro and in vivo conditions	60
	Filamentous phages used for delivery of Chloramphenicol prodrug effectively retarded growth of *Staphylococcus aureus*	61
Expanding host range	Adding receptor binding domain from IKe g3p phage to the end of fd g3p bacteriophage enables the latter to infect bacteria bearing both N and F pili	54
	Modification of tail fiber protein (gp38) enables T2 phage to infect *Escherichia coli* O157:H7	62
Enhancing antimicrobial activity	Engineered phage carrying the *lexA3* gene acts as an antibiotic adjuvant by suppressing the SOS response and increased *Escherichia coli* cell death	63
Alteration of lysogenic to lytic phage	Deletion of vir gene converted lysogenic bacteriophage to lytic one	64
Production of lethal but nonlytic compound	The genetically engineered M13 phage encoding restriction endonuclease *BglII* gene or modified phage λS holin genes efficiently killed *Escherichia coli* while leaving the cells structurally intact	65

host range. Table 5.2 shows some possibilities of biofilm control by naturally and genetically modified phages.

Application of phage cocktail is another strategy for overcoming the limitation of narrow host range. Ganegama Arachchi et al.[66] investigated the effect of three bacteriophages (individually or in combination) on *Listeria monocytogenes* biofilm formed on stainless steel coupons. Although each individual phage reduced the sessile cell counts by 3–4.5 logs, the phage cocktail reduced the sessile cells to an undetectable level. Phage cocktail can inhibit biofilm for a longer period of time than single phages.[67] For the development of efficient phage cocktail, a proper understanding of phage host range is needed. It should be noted that the host range determination is a costly and time-consuming process. Also, host range is not a fixed property and phages can change their specificity over time.[68,69]

Lysogeny

Lysogenic phages are concerning problem in phage treatment of MIC because the transition of lytic to lysogenic phages increases the chance of bacterial survival. Meader et al.[70] investigated the effect of bacteriophage ΦCD27 on *Clostridium diffcile* population. This bacteriophage significantly reduced the bacteria in two of the three replicates. However, in one replicate, there was no significant difference in bacterial number compared with the control. Upon exposure of bacteria in this vessel to mitomycin C, which is known to cause induction of temperate phages, active phage particles were released. Because of transduction potential, lysogenic phage may spread the antibiotic resistance gene and, therefore, develop more resistant bacteria. Lysogenic phage may enhance the formation of bacterial biofilm. Gödeke et al.[71] observed that the prophage-mediated lysis of *Shewanella oneidensis* MR-1 can induce the release of crucial biofilm-promoting factors, in particular, eDNA. It has been shown that a phage cocktail with a proper combination of phages can effectively prevent the lysogenic bacterial colon.[72] Use of lysogenic phage, which is expressing bacteriocidal agents without attacking the bacterial cell, is another strategy to overcome this problem.[65] To study the effect of phage therapy on *Clostridium difficile* infection, Nale et al.[72] conducted a study and determined the host range of seven bacteriophages on 80 strains of *C. difficile*. The examined phages were able to lyse 62 different strains of bacteria. They observed that although treatment of some strains with a single phage resulted in the appearance of phage-resistant colonies, treatment of colonies with specific combinations of phages prevented the emergence of resistant clones as well as lysogenic one. Using an animal model for study, they also observed that the proper combination of phages reduced the bacterial colonization and delayed the onset of symptoms. Rapson et al.[64] engineered a bacteriophage by deleting its *vir* gene. Mutation of this gene resulted in the permanent conversion of lysogenic phages to the lytic one.

Bacterial resistance

Another challenge during phage treatment is resistant to phages. Unlike resistance to predatory bacteria, resistance to phages is a mutational event and it occurs within few hours in vitro. The mechanisms of resistance to phages including but not limited to (1) preentry resistance mechanisms such as loss or lack of receptor, structural modification, and/or masking of the receptor and (2) postentry resistance mechanisms such as degradation of phage DNA by restriction-modification defense systems and abortive infection systems, which limit viral replication.[73]

Development of resistance against phages decreases the efficiency of phage treatment. Survival of the resistant bacteria may act as a root for reestablishment of biofilm. Therefore for successful application of phage treatment, the rate of bacterial immunity to phages should be reduced. Phage cocktail can be exploited for this purpose. Fu et al.[67] observed that the pretreatment of a catheter with a five-phage cocktail reduced the biofilm cell density of *Pseudomonas aeruginosa* by 99.9%. In addition, exposure of biofilm to the phage cocktail resulted in lower resistant of bacteria to the phages.

Disruption of resistance mechanisms is another strategy to reduce bacterial resistance. Bacterial capsule and EPS impede viral access to bacterial cell surface. By using fluorescence correlation spectroscopy with two-photon excitation, Briandet et al.[74] quantified the dispersion of lactococcal c2 phage through model diffusion and observed that the phage can penetrate inside biofilm through water channels. Although they mentioned that the EPS is not restricting the phage penetration, many other studies have shown that phage penetration is facilitating when EPS is degraded by phage enzymes such as polysaccharide lyases.[75] Lu and Collins[55] engineered an *Escherichia coli* bacteriophage called T7 to express dispersin B, a hydrolase enzyme known to have biofilm-dispersing activity. They observed that compared with parent T7, which was lacking

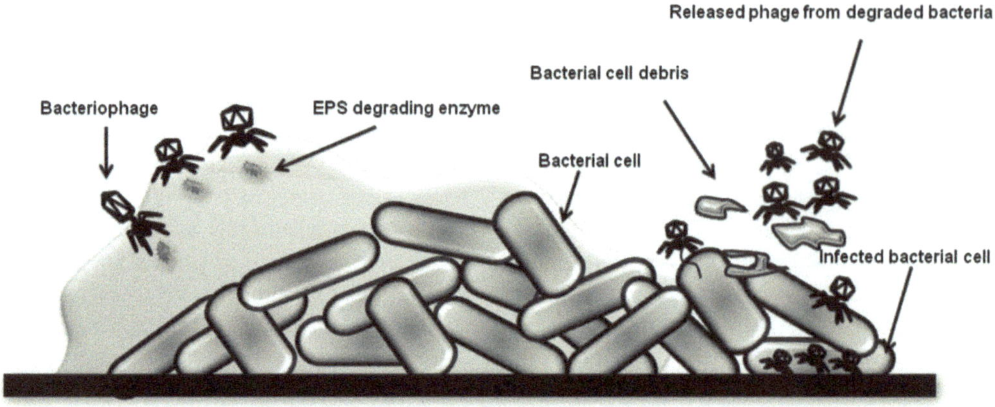

FIG. 5.5 Schematic representation of corrosion inhibition mechanism by phage.

the enzyme, the recombinant phages reduced the bacterial biofilm by more than 10^4 fold. Phages with exopolymerase enzymes can attribute to biofilm reduction not only by eradication of bacteria but also by degradation of the biofilm matrix (Fig. 5.5). These phages are having one more advantage. They avoid the pseudolysogeny by preventing phage trapping in the biofilm.[69] Pseudolysogeny postpones the release of phages and substantially reduces the treatment efficacy. It is important to note that the enzyme-producing phages do not necessarily secrete it under all conditions.[76]

Because the EPS complexity is lower than that of the host bacteria, it has been suggested that the use of phage products such as EPS-degrading enzyme could be more effective than implementing the entire phage particle for removal of biofilm. This approach has other advantages. Use of phage product reduces the phage resistant in bacteria and also eliminates the chance of transmitting destructive properties to bacteria through transduction. However, the hypothesis about the use of phage products was rejected when the activity of *Staphylococcus aureus* bacteriophage virion was compared with its depolymerase enzyme on multispecies biofilm. None of them affected bacteria other than *S. aureus*.[77] EPS-degrading enzymes, in general, have narrow substrate specificity. These enzymes are very specific and seldom act on more than a few closely related polysaccharide structures.[78] Each enzyme at most degrades a few related polysaccharides, which constitute a small proportion of pool of polysaccharides in natural multispecies biofilm. This process just increases the patchy and heterogeneous structure of biofilm and definitely increases the corrosion rate. Therefore, multiple depolymerases would be required for targeting mixed biofilm. In addition to specificity, the high cost of enzyme also is a drawback for application of phage products. These enzymes are not stable in all conditions. High pH, temperature, and salt concentrations could denature them. Therefore, while using EPS-degrading enzymes (or virions expressing EPS-degrading enzymes), operation condition should be considered.[53] Phages expressing enzymes with broad antiquorum sensing activity are a good substitute for phages with depolymerases enzyme because the former can overcome both substrate and host specificity. An engineered T7 phage, which encodes a lactonase enzyme, inhibits the formation of a mixed-species biofilm of *Pseudomonas aeruginosa* and *Escherichia coli*[59]. Lactonase is an enzyme with a broad-range activity that can quench the activity of acyl homoserine lactones (AHLs).

Although phage-based therapeutic approaches have their own limitations, they have some advantages that make them more appropriate candidate than the biocidal agents. For instance, phages are able to infect persister cells.[79] Pearl et al.[80] reported that the phages could replicate inside the persister cells if they become metabolically

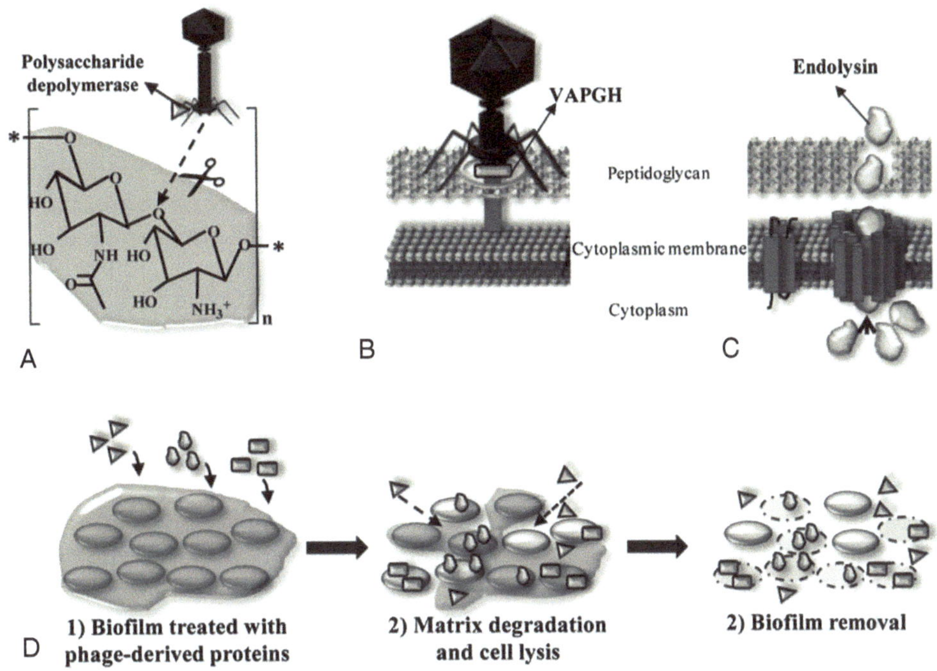

FIG. 5.6 Model shows how exopolysaccharide depolymerase, VAPGH, and endolysin enzyme degrade bacterial biofilm: **(A)** exopolysaccharide depolymerase enzyme cleaves the glycosidic bonds of extracellular matrix, **(B)** VAPGH is making a hole in the bacterial cell wall, **(C)** endolysin mediates bacterial lysis at the end of infection cycle, and **(D)** antibiofilm activities of exogenous polysaccharide depolymerase, VAPGH, and endolysin enzyme. (Data obtained from GutiérrezD, Ruas-Madiedo P, Martínez B, Rodríguez A and García P, Effective removal of *Staphylococcal* biofilms by the endolysin LysH5. *PLoS One*. 2014;9(9):e107307.)

active. Because bacteria located on the surface of biofilm are metabolically active and are able to support larger phage burst than the cells deeper in the biofilm, phages are more effectively infecting bacteria found in biofilm peripheries than cells deeper in the biofilm.[81,82] Therefore, it is more reasonable to use phages for pretreatment of biofilm or eradication of young biofilm. The limitation of phages for the treatment of mature biofilm can be circumvented by making biofilm-interior bacteria more surface located or by combining phages with predatory bacteria, which attack in depth bacteria. Another strategy to overcome this limitation is the use of phages encoding lytic protein such as endolysins and virion-associated peptidoglycan hydrolyses enzymes (VAPGHs). (Endolysin is a hydrolytic enzyme produced by bacteriophages at the end of a lytic cycle. The endolysin facilitates the release of progenies. VAPGH is a hydrolytic enzyme produced at the beginning of a lytic cycle. This enzyme makes a hole through which the phage tail crosses the host cell wall and then the phage injects its genetic material into host cell cytoplasm.)[83] These enzymes degrade the peptidoglycan and destabilize the bacterial cell wall (Fig. 5.6). It is interesting to know that the spectrum activity of endolysin is broader than the host range of bacteriophages and no phage-resistant bacteria to this protein has been reported to date.[83,84] However, endolysin has one drawback. This enzyme is ineffective on Gram-negative bacteria as their outer membrane, which is impermeable to this enzyme, protects the layer of peptidoglycan.[85] Shen et al.[86] evaluated the effect of a traditional antibiotic as well as engineered bacteriophage encoding endolysin on both metabolically active and inactive cells of group A *Streptococci* in the biofilm. Laser scanning confocal microscopy showed that the recombinant phages easily penetrate inside biofilm and destroy the bacterial biofilm. Although the effect of antibiotic and recombinant phages was similar on the planktonic bacteria, the lysis effect of recombinant phages was more effective than the traditional antibiotic. Application of phages that are suppressing SOS response network (SOS is a cell response to DNA damage, which results in DNA repair) in bacteria is another alternative to treat mature biofilm. It has been shown that these bacteriophages can eliminate persisters, antibiotic-resistant bacteria, and biofilm cells.[63]

The attention devoted to the variables affecting phage-mediated infection can improve the removal of bacterial biofilm.[87] For instance, as smaller phages penetrate more easily into the biofilm matrix and produce more progenies, it is better to use them for inhibition of biofilm.[88] In a study conducted by Soni et al.,[89] they observed that the 30 and 15 min contact time with phages results in more or less than 1 log CFU/g reduction in the *Listeria monocytogenes*, respectively. Appropriate use of phage-to-host ratio or multiplicity of infection (MOI) is another factor that influences the treatment process. Hanlon et al.[90] observed that the use of 100:1 MOI reduces the *Pseudomonas aeruginosa* biofilm by 1 log, whereas 1000:1 MOI reduces the biofilm by 2 logs. Similarly, Knezevic and Petrovic[91] demonstrated the ability of phage л-4 for the elimination of *Pseudomonas aeruginosa* PA-4u. At MOI of 10^1, biofilm was reduced by more than 50%, and at MOI of 10^3, the phage eliminated the biofilm completely. Fernandez et al.[92] showed that the treatment of biofilm by a nonlethal dose of phages not only can result in treatment failure but also can promote the formation of thicker biofilm with greater DNA content.

In the event of a real-life application of phage, it is necessary to consider the effect of environmental conditions on the phage application. The survival and persistence of phages are known to be affected by various biotic (host morphology) and abiotic factors (temperature, pH, salinity).[87] Phage virions do not move independently. Several environmental factors such as flow rate, pH, ionic strength, and presence of organic matter not only may influence the speed of phage diffusion but also may affect the phage adsorption.[93] Phage attachment to any improper receptor reduces the phage treatment efficacy.[94]

Phage coating and immobilization of phages on the metal surface are two strategies, which could be used for removal of these obstacles. Recently, Anany et al.[95] immobilized *Listeria monocytogenes* and *Escherichia coli* 0157:H7 phages on the modified cellulose membrane and observed that these phages were effectively controlling the growth of bacteria. Li et al.[96] observed that the conjugation of polyvalent phage called PEL1 with magnetic nanoparticles facilitates its migration under a relatively small magnetic field. Because conjugated phages could penetrate deeper layers of biofilm, they efficiently eradicated the biofilm.

Method of application is another factor which should be taken into account. Several studies have shown that the combination of phages and bactericidal agents is several orders of magnitude more efficacious than any of these items alone.[92,97,98] Chaudhry et al.[99] assessed the independent and synergistic effect of two phages and five bacteriocidal antibiotics on biofilm of *Pseudomonas aeruginosa*. The result showed that the combination of antibiotics and phages is more effective than the treatment with any of the individual agents. Interestingly, during the combination of phage treatment and antibiotic therapy, some antibiotics were showing better result at the lower concentration than the higher one. This result was contradictory to the research of Fernandez and his colleagues.[92] While studying the synergistic influence of mupirocin antibiotic and phiIPLA-RODI phages on biofilm of *Staphylococcus aureus* IPLA1, Fernandez et al.[92] observed that the subinhibitory concentration of mupirocin increased the resistance of *S. aureus* to respective bacteriophages. Therefore the concentration of biocide is a critical factor, which should be considered during combination therapy. Time of phage delivery is another important factor, and it varies based on coadministrated strategies. For instance, phage addition should be done after mechanical disruption of biofilm because phages more effectively infect planktonic bacteria than the sessile one.[66] However, it is more practical to do phage treatment before chemical treatment as disruption of biofilm by phage therapeutic strategy enhances bacterial switch from sessile to planktonic phase and improves the effectiveness of biocide treatment.

CORROSION INHIBITION BY NITRATE-REDUCING BACTERIA

Nitrate is one of the most important nutrients for microbes. Microorganisms uptake nitrate through either an assimilatory or dissimilatory pathway. In the assimilatory process, nitrate is incorporated into microbial biomass, whereas in dissimilatory pathway, nitrate is used as an electron acceptor for the aim of energy production. In the dissimilatory pathway, nitrate reduction goes either through denitrification where nitrate converts to nitrogen gas or it proceeds via dissimilatory nitrate reduction to ammonium (DNRA) pathway where nitrate conversion to ammonia occurs (Fig. 5.7). Bacteria with the ability to perform denitrification are called denitrifying bacteria or nitrate-reducing bacteria (NRB).[100] Denitrifiers have the ability to suppress some of the most important microbial activities involved in MIC. It has been shown that the nitrite produced by the nitrifying activity of *Shewanella putrefaciens* 200 had an inhibitory effect on electron transport to Fe^{3+}.[101] NRB inhibits the methanogenesis either by capturing the hydrogen normally converted to methane or by producing intermediates such as nitrite, nitric oxide, and nitrous oxide, which directly target the methanogenic enzymes[102] (Fig. 5.8B). The intermediates formed during denitrification can also suppress the sulfate reduction activity of bacteria.[103,104] Nitrite is inhibitory to dissimilatory sulfite reductase, an enzyme responsible for the reduction of sulfite to sulfide. Nitrous oxide disturbs the redox balance in a given environment to such an extent that strict anaerobic bacteria like SRB get inhibited.[105] In addition to the abovementioned mechanisms, different groups of NRB (i.e., nitrate-reducing–sulfide-oxidizing bacteria and heterotrophic nitrate-reducing bacteria) can exclude SRB by their specific mechanisms. Heterotrophic nitrate-reducing bacteria (hNRB) compete with SRB for degradable organics carbon (Fig. 5.8A). So hNRB reduce the amount of available food and limit the growth of SRB. Grigoryan et al.[106] observed that SRB and hNRB are able to use similar sources of electron donor but in different ways. In microcosm containing three different volatile fatty acids (acetate, propionate, and butyrate), SRB are initially converting the propionate and butyrate into acetate, and once the concentration of acetate reaches the critical level, then SRB use it as an electron donor. However, hNRB could consume all three compounds simultaneously. In a competition for the same electron donor, hNRB will overcome SRB because, for a

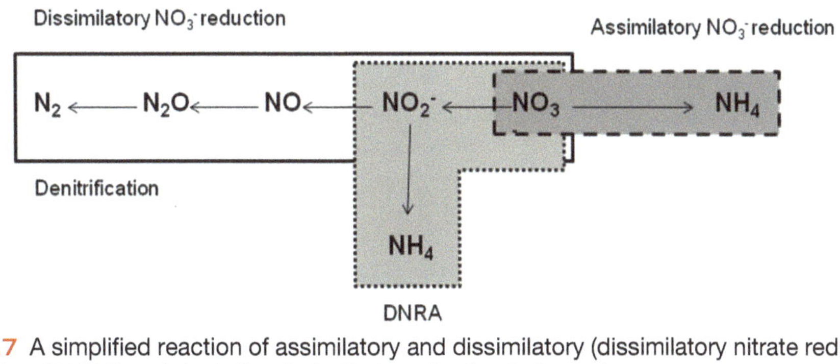

FIG. 5.7 A simplified reaction of assimilatory and dissimilatory (dissimilatory nitrate reduction to ammonium [DNRA] and denitrification) nitrate reduction in bacteria.

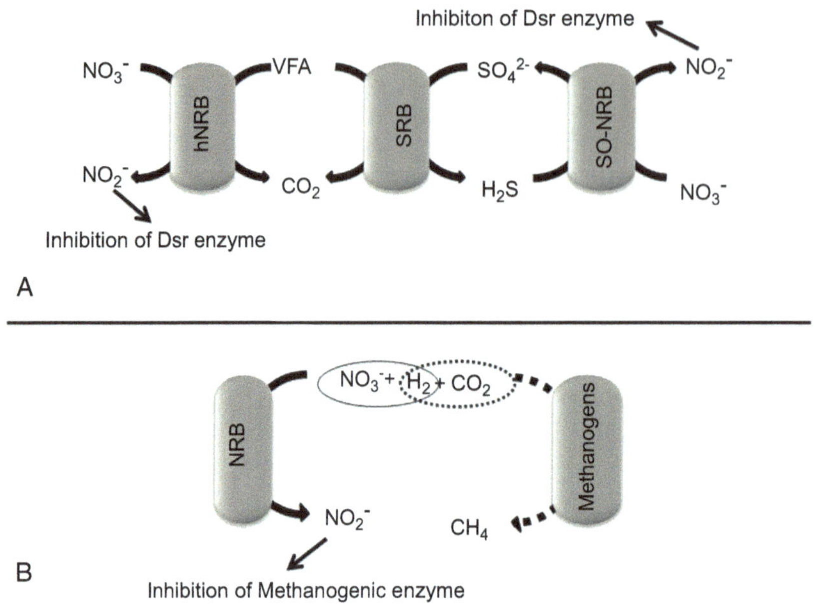

FIG. 5.8 Schematic representation of corrosion inhibition mechanisms by nitrate-reducing bacteria. **(A)** Exclusion of sulfate-reducing bacteria and **(B)** methanogens by NRB.

given electron donor, the energy gained from nitrate reduction is greater than the energy obtained from sulfate reduction as shown below.[107]

$$5CH_3CO_2^- + 8NO_3^- + 3H^+ \rightarrow 10HCO_3^- + 4N_2 + 4H_2O$$

$$\Delta G = -495\ kJ\ (mol\ NO_3^-)^{-1}$$

$$CH_3CO_2^- + SO_4^{2-} \rightarrow 2HCO_3^- + HS^-$$

$$\Delta G = -47\ kJ\ (mol\ SO_4^{2-})^{-1}$$

In contrast to hNRB, nitrate-reducing–sulfide-oxidizing bacteria (NR-SOB) increase the redox potential of an environment by oxidizing the hydrogen sulfide[108] (Fig. 5.8A).

Theoretically, by elimination of SRB and inhibition of hydrogen sulfide production, corrosion rate can be reduced, but in practice, it is more complicated. There are several studies that show control of SRB via nitrate treatment does not necessarily reduce corrosion or souring risks.[109–111] For instance, when Schwermer et al.[111] studied the effect of nitrate treatment on pitting corrosion of pipelines caused by SRB, they observed that the pitting corrosion was not affected by the treatment, whereas uniform corrosion was reduced to a great extent. This occurred while addition of nitrate changed the bacterial community from high-diversity indigenous bacterial population into low-diversity NRB population (which led to a significant increase in nitrate reduction in comparison with dissimilatory sulfate reduction) and also when conversion of nitrate to nitrite occurred in the biofilm adjacent to the metal surface. It is believed that pitting corrosion was mediated by nitrite. Nitrite is known as anodic inhibitor; however, an insufficient concentration of it may cause pitting corrosion.[112]

Other studies also demonstrated that the NRB may cause corrosion acceleration. Yin and Wunch[110] investigated the effect of nitrate treatment on corrosion of carbon steel in reactors containing SRB and either nitrite-producing NRB or NR-SOB. As compared with the control bioreactor, which did not receive nitrate, in both reactors the amount of microbiologically produced sulfide has reduced and corrosion of carbon steel has increased. Authors mentioned that in the reactor with nitrite-producing NRB, biologically produced nitrite reacted with the remaining sulfide and generated highly corrosive polysulfide compound. In the second reactor, the elemental sulfur produced because of sulfide-oxidizing activity of NR-SOB, was aggressive toward the carbon steel. Therefore, attention to the NRB metabolite is critical for successful implementation of nitrate.

The chance of treatment failure will increase if SRB with nitrate-reducing activity dominates in the reservoir. Some SRB such as *Desulfovibrio vulgaris* Hildenbourgh can use nitrate as an electron acceptor. In these bacteria, nitrite reductase enzyme reduces nitrite to ammonia and inhibits nitrite entry into dissimilatory sulfite reductase enzyme.[113] In this case, nitrate addition could be problematic as it leads to silent souring. Nitrate stimulates SRB growth and SRB could outburst when nitrate concentration reduces in the reservoir. Therefore, it is of particular importance to monitor the presence of SRB during the time of nitrate treatment. However, note that SRB activity should not be evaluated by analyzing the amount of sulfide produced or sulfate reduced, as ammonification activity of SRB can lead to a false understanding of the number of SRB.

Presence of SRB that acts as a direct catalyst of corrosion also should be considered during the treatment process. Under carbon starvation, such SRB can highly increase corrosion by direct uptake of electrons from metal coupons.[114] Therefore in the circumstances where nitrate addition leads to the outbreak of hNRB, these SRB will suffer from carbon source limitation and therefore they switch to elemental iron as the electron donor, which substantially increases the rate of corrosion. Direct electron uptake from metal has also been observed in nitrate-reducing bacteria. Xu et al.[115] demonstrated that during the carbon starvation, biofilm of nitrate-reducing *Bacillus licheniformis* grown on carbon steel triggers more aggressive corrosion than *Desulfovibrio vulgaris*. Thus the availability of carbon source during the treatment is essential for preventing sever MIC. During nitrate treatment, the role of electron transfer mediators should be considered as these compounds can enhance the extracellular electron transfer in bacteria, which subsequently results in acceleration of corrosion. Jia et al.[116] reported that the addition of either 10 ppm riboflavin or 10 ppm flavin adenine dinucleotide (FAD) to the culture media of nitrate-reducing *Pseudomonas aeruginosa* PAO1 has no effect on the planktonic cell counts; however, it significantly can enhance the maximum pit depth on carbon steel coupons from 17.5 to 24.4 μm and 25.0 μm, respectively. Addition of these compounds also increased the specific weight loss of carbon steel from 2.06 to 2.34 mg/cm.

It has been demonstrated that the efficient application of nitrate treatment relies on microbial composition and their interaction in oil reservoirs. Both nitrite-producing bacteria (NO_2NRB) and denitrifiers (N_2NRB) are able to eliminate the SRB population. However, the coexistence of these two NRB reduces the treatment efficacy because N_2NRB exclude the nitrite produced by NO_2NRB.[110] Therefore in a reservoir containing both NO_2NRB and N_2NRB, nitrate should be added when the NO_2NRB are the dominant population.

Monitoring the end products of nitrate reduction also is important for the successful operation of nitrate treatment. The ratio of nitrate/sulfide and also the ratio of nitrate/organic carbon determine the fate of nitrate (or nitrite). At high nitrate to sulfide ratio, *Sulfurimonas* sp. strain CVO converts nitrate and sulfide into nitrite and sulfate, whereas these compounds get converted into nitrogen and sulfur at low nitrate to sulfide ratio.[117] *Sulfurospirillum* sp. strain KW undergoes denitrification at low C/N ratio and ammonification at high C/N ratios.[104] Conversion of nitrate to ammonia, which does not have any inhibitory effect on SRB, depletes nitrate from oil field systems and increases the need for higher concentration of nitrate. Other factors such as temperature and pH also influence the destiny of nitrate reduction. For instance, cultivation of *Shewanella loihica* strain PV-4, an iron-reducing bacterium with nitrate reduction ability at 30°C or pH above 7.0, will support the ammonification.[118]

Erşan et al.[119] reported that the ratio of aggressive to protective ions affected corrosion control by NRB. They examined the effect of nonaxenic (bacterial culture with more than one species) NRB with $CaCO_3$ precipitating ability on simultaneous concrete crack healing and corrosion inhibition of inside rebar in a corrosive electrolyte solution (0.05 M NaCl). In the presence of calcium ions, NRB activity led to calcium carbonate precipitation and subsequent repair of concrete cracks. Besides, production of nitrite and its reaction with the ferrous ion led to passivation of rebar surface. However, stability of protective passive layer was depended on nitrite to chlorine ion concentration as aggressive Cl^- ion could compete with nitrite for reacting with Fe^{2+}.[119]

Method of application is a key factor in successful corrosion control by NRB. For example, as hNRB is inhibited by high concentration of sulfide, it is reasonable to reduce the concentration of sulfide in oil field through the action of NR-SOB and then stimulate the growth of hNRB.[120] Employment of bacteria with both hNRB and NR-SOB activity such as *Sulfurospirillum* sp. is another alternative to overcome high sulfide and low carbon concentration.[104] As nitrite reductase has never been demonstrated in thermophilic SRB, nitrite may be a suitable inhibitor for sulfate reduction reaction in high-temperature oil fields.[121]

Nitrate injection is recommended in oil fields containing hNRB and NR-SOB because nitrate has a stimulatory effect on NRB growth, whereas injection of nitrite is more appropriate into fields containing SRB and no NRB because nitrite directly inhibits SRB. The injection dose depends on the concentration of sulfide and organic carbon as the ratio of nitrate to these compounds determines the bacterial product.[122] Instead of a continuous injection, pulse injection of nitrate should be done because it will prevent the succession of the microbial community.[123] Voordouw et al.[124] observed recovery of souring at low-temperature oil reservoir after continuous addition of 2 mM nitrate to injection water. They found that the recovery was due to zonation of the microbial community in which NRB grew close to the injection wellbore, whereas SRB was presented deeper in the reservoir.

In spite of all limitations mentioned above, there are several reports about the successful industrial application of nitrate treatment. Substitution of biocide by nitrate during the treatment of injection water of Gullfaks platform led to 1000-fold reduction in SRB number, 10%–20% reduction in sulfide production, and 50% reduction in corrosion rate.[125] Long-term nitrate treatment (2002–07) on corrosion control was more effective as compared with biocide treatment in Gullfaks and Veslefrikk oilfields. Injection of biocide in these oil fields stimulated SRB activity and increased rate of corrosion, whereas nitrate injection reduced the activity of SRB in both oil fields and increased the NRB number. Although nitrate treatment led to 40% corrosion reduction at Gullfaks field, it did not significantly reduce the rate of corrosion at Veslefrikk oilfield. However, corrosion rate is found to be more stable at Veslefrikk during the nitrate treatment as compared with corrosion rate during the biocide treatment.[126]

CORROSION INHIBITION BY REGENERATIVE BIOFILM

Bacterial biofilm has corrosion inhibitory potential. Biofilm can coat the metal surface and isolate it from the surrounding corrosive environment. Exploitation of biofilm for corrosion control has several advantages such as production of antimicrobial (or antibiofilm/anticorrosive) compounds from inside the biofilm, which results in availability of higher concentration of antimicrobial in the biofilm.

The biofilm that acts as an inhibitor film protects metal from corrosion by providing a barrier between the aggressive environment and bare metal. Therefore, biofilm prevents the charge transfer between the anode and cathode.[127] In spite of the excellent characteristics a film might possess, its inhibition ability is highly dependent on the adhesion to the surface.[128] Jayaraman et al.[129] showed that a good biofilm-forming bacterium, *Pseudomonas putida*, can greatly inhibit the corrosion of carbon steel in LB medium, while a poor biofilm-forming bacterium, *Streptomyces lividans*, was not able to inhibit the corrosion of same metal in LB medium effectively. Abdoli et al.[130] studied the corrosion behavior of thermal sprayed Al and Al/nano-Al_2O composite coatings in artificial seawater in the presence of *Escherichia coli* bacteria. *E. coli* biofilm formed after 7 days of incubation. Initially, the bacterial biofilm was patchy and discontinuous while longer incubation resulted in more bacterial colonization and formation of homogeneous biofilm. Electrochemical study of corrosion has revealed that the corrosion inhibition has increased when homogenous biofilm covered the full surfaces of metal.

Several different factors are affecting the adhesion properties of biofilm.[131] Therefore, while applying regenerative biofilm, role of these factors should be considered. Biofilm composition is one of these factors, which contributes in both strength and cohesiveness of biofilm. Polysaccharides and proteins in biofilm enhance bacterial adhesion, while lipopolysaccharides, uronic acids, and biosurfactants discourage adhesion.[131–135] Structure of these compounds also influences the cohesiveness of biofilm. Adhesion of bacteria having longer lipopolysaccharides (LPS) to the metal surface is more strong compared with the one having shorter LPS.[136] While studying the effect of two extracted extracellular polymeric substances isolated from two different strains of *Leuconostoc mesenteroides* on corrosion of low-carbon steel, Finkenstadt et al.[137] noticed that some unknown factor(s) play a role in adhesion of bacteria to the steel surface. One of these EPS immediately reacted with the steel substrate and formed a passive layer, whereas another did not. Because polysaccharides with a similar structure did not show the same effect on corrosion protection, it was concluded that corrosion inhibition was strain specific.

The surface property of the metal (metal surface roughness, polarization, oxide coverage, and chemical composition) is another factor that influences the bacterial adhesion. This effect can be clearly seen from the result of the study of Juzeliunas et al.[138]. They investigated the corrosion inhibition effect of *Bacillus mycoides* on aluminum, zinc, and mild steel in a 2-year study under controlled conditions (humidity and temperature). Although the environmental condition was identical in all experiments, bacteria accelerated corrosion of zinc, inhibited corrosion of aluminum, and showed no effect on mild steel corrosion. Li and Logan[136] studied adhesion of seven Gram-negative and one Gram-positive bacteria to 11 metal oxide–coated and uncoated glass surfaces. They found a significant increase in bacterial adhesion to glass surfaces covered with metal oxide, as compared with uncoated glass. Under the same ionic strength, the enhanced attachment was attributed to increased surface roughness and surface hydrophobicity but not to surface charge.

Not only the film attachment but also its homogeneity affects the inhibitory properties. Wadood et al.[139] studied the corrosion of SS304 in minimal salt medium supplemented with 1.5% NaCl in presence of single- and mixed-species biofilm of *Bacillus subtilis* S1X and *Pseudomonas aeruginosa* ZK. They observed that both single-species biofilm of *B. subtilis* and *P. aeruginosa* increased the corrosion resistance of metal. Corrosion protection was higher in presence of *B. subtilis.* Authors have mentioned that the higher efficiency of *B. subtilis* is related to its biofilm structure. This bacterium produced a thick and compact biofilm, which prevented the transfer of corrosive ion to the metal surface. Because single-species biofilms were showing corrosion inhibitory properties, the expectation was for a strong inhibitory result by dual-species biofilm. However, exposure of metal to the mixed biofilm gave rise in moderate corrosion inhibition as coculture of these bacteria resulted in the formation of more heterogeneous biofilm. Presence of different metabolic types of microorganisms, which surely interact with each other through the variety of physiologic and genetical processes, may deviate the protective function of inhibitory biofilm. Local variation of microbial substrate and products in biofilm increases the biofilm heterogeneity as well. Formation of chemical gradient in biofilm not only creates microniches in the biofilm, which supports the growth of different microorganisms in biofilm, but also allows the formation of differential concentration cells.

It is important to note that an inhibitory film is protective if it remains intact. Similarly, any destruction of protective biofilm accelerates the corrosion rate. The immunity of the biofilm to attack is a function of the biofilm composition, its density, and thickness. Biofilms formed in anoxic condition are thinner, denser, and more uniform than biofilms formed in aerobic condition.[140,141] The thinner and denser anoxic biofilms are more susceptible to erosion, whereas thicker aerobic biofilms are more prone to sloughing. Sloughing and erosion alter the biofilm structure and make it patchy and discontinuous, which accelerates corrosion of metal through the formation of differential concentration cells. Therefore, to take advantages of regenerative biofilms, their behavior must be controlled. Although several studies have been performed and their results have opened new insights into successful industrialization of this type of biofilm, so far no bacterial biofilm has successfully protected the metal from corroding.

CORROSION INHIBITION THROUGH THE REMOVAL OF CORROSIVE AGENTS

Bacterial biofilm acts as a barrier and limits the diffusion of material. It has been shown that a biofilm of 12 μm thickness reduces the concentration of oxygen at the base of a biofilms, whereas a biofilm with 300–400 μm thickness inhibits penetration of sulfate.[142,143] However, when Ismail et al.[144] observed that the live viable bacterial biofilm is necessary for corrosion inhibition, it was determined that biofilm is not just a barrier and it is actively playing a role in corrosion inhibition. Jayaraman et al.[145] observed that the corrosion of SAE 1018s steel has reduced in presence of biofilm-forming bacteria (*Pseudomonas fragi* and *Escherichia coli DH5a)* grown in different culture media (i.e., Luria-Bertani, Vätäänen nine-salt solution, and modified Baar's medium) in both batch and continuous culture conditions. Because corrosion rate of steel exposed to these biofilms was similar to corrosion rate of steel exposed to anaerobic sterile medium and also anaerobically grown *E. coli* DH5a and it was lower than the rate of corrosion of steel placed in the aerobic sterile medium, they mentioned that corrosion inhibition was mediated through oxygen removal. It is important to note that production of hydrogen peroxide, a compound with higher redox potential than oxygen, during the bacterial respiratory activity may result in corrosion acceleration. Although bacteria can decompose the hydrogen peroxide into oxygen and water by enzymatic activity,[146] Busalmen et al.[147] observed that catalase containing *Pseudomonas* sp. can reduce the newly formed oxygen to hydrogen peroxide again and increase the cathodic current. Karn et al.[148] showed that the correlation between catalase and peroxidase activity and corrosion is complicated. They studied the potential effect of peroxidase and catalase enzyme produced by five different *Bacillus* species (designated as DS1 to DS5) on corrosion of carbon steel. Although the high concentration of enzymes has found in the biofilm of all bacteria, the concentration of enzymes varied among species. There was a direct link between production of peroxidase and corrosion inhibition (DS1 and DS3, which induced the corrosion, produced 22.18 and 19.54 U/mL peroxidase, respectively; production of peroxidase for DS2 and DS5 with no inhibitory effect on corrosion was 28.57 and 27.0 U/mL and DS4 with inhibitory effect was able to produce 36 U/mL peroxidase). However, there was no clear relation between catalase production and corrosion. The highest catalase production was observed in DS1 (67.28 U/mL) and DS3 (61.57 U/mL), the bacteria which were inducing the corrosion. The catalase concentration in the biofilm of DS2 and DS5 bacteria, which did not have any inhibitory effect on corrosion, was 57.71 and 59.14 U/mL and it was 70.14 U/mL in the biofilm of DS4 with inhibitory effect.

Dubiel et al.[149] reported a decrease in corrosion because of the reaction of bacterial metabolite with corrosive oxygen. In anaerobic condition, *Shewanella oneidensis* uses Fe (III) as a terminal electron acceptor and reduces it to Fe (II). Fe (II) reacts with the oxygen and excludes oxygen from the bulk fluid. There are other mechanisms by which microorganisms can reduce the concentration of the cathodic reactant at the metal surface. Soleimani et al.[150] investigated the effectiveness of *Escherichia coli* DH5α biofilm on membrane protection against hydrogen sulfide by using a sulfide biosensor. They observed that *Escherichia coli* biofilm blocked the penetration of corrosive sulfide ions and therefore it helped in reducing MIC.

CORROSION INHIBITION THROUGH A PROTECTIVE LAYER

Formation of passive layer on metal surface results in corrosion reduction. Microbes can contribute to corrosion resistance by reinforcement of passive film. Chongdar et al.[151] showed that the formation of bacterial EPS on a chemically formed passive layer can increase the corrosion protection of metal surfaces. They studied the influence of two different aerobic bacteria, viz., *Pseudomonas alcaligenes* and *Pseudomonas cichorii*, on the corrosion of mild steel in phosphate-buffered basal salt solution. They observed that the inhibition of corrosion by *P. cichorii* was higher than the corrosion inhibition observed in the presence of *P. alcaligenes*. Examination of mild steel surfaces showed that the steel surface exposed to *P. alcaligenes* was covered with iron oxide and iron phosphate and it was covered by iron-EPS complex when exposed to *P. cichorii*.

Microbes can also contribute to corrosion inhibition by producing metabolites, which accelerate the metal passivation. In mineral medium containing more than 2 mM phosphate, *Rhodococcus* sp. strain C125 and *Pseudomonas putida* mt2 form insoluble iron (II) phosphate layer (vivianite layer) on the mild steel and inhibit corrosion of the steel.[152] In siderophore availability, *Pseudomonas putida* can even restore the damaged vivianite layer.[153] Siderophores are microbial metabolites that assist in passivation of metals. Siderophores bind to diffused metal cation and form a metal-siderophore complex. This complex will precipitate near the metal surface and forms a protective layer on it. Affinity and stability of siderophore-metal complex determine the efficiency of passive film, which is strongly affected by the environmental factors such as pH, metal concentration, and composition.[154,155]

CORROSION INHIBITION THROUGH ANTIMICROBIAL PRODUCING BIOFILM

Biofilm can be used for production of antimicrobial agents (Fig. 5.9). Biofilm with antimicrobial activity can potentially be as effective as biocide with some distinct advantages. By in situ production of antimicrobials within biofilm, not only the problem of diffusion limitation within biofilm will be solved but also a high concentration of antimicrobial compound will be maintained inside the biofilm.[156] For the first time, Jayaraman et al.[156] studied the effect of antimicrobial-producing biofilm for combating MIC. They observed that in situ production of antimicrobial indolicidin, bactenecin, and probactenecin by genetically engineered *Bacillus subtilis* resulted in significant inhibition of SRB growth and subsequently MIC inhibition of mild steel. Zuo and Wood[157] investigated the effect of gramicidin S–producing biofilms of *Bacillus brevis* on microbial corrosion of mild steel in presence of two bacteria; sulfate-reducing *Desulfosporosinus orientis;* and

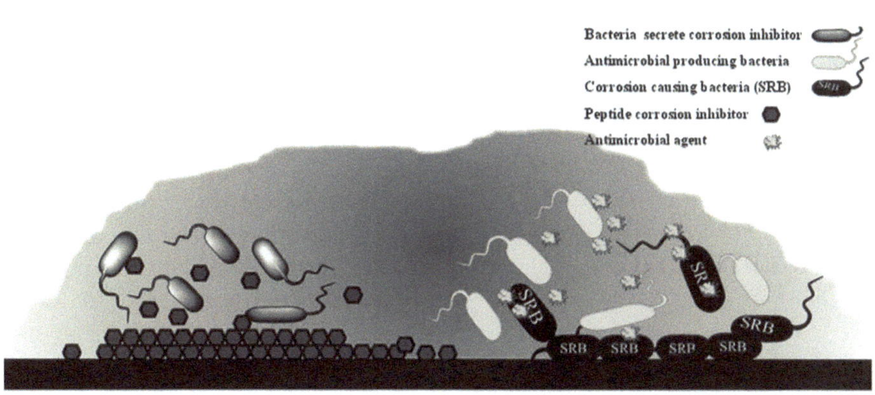

FIG. 5.9 Schematic representation of corrosion inhibition by bacteria producing corrosion inhibitor and antimicrobial compounds. (Data from Alasvand Zarasvand K and Rai VR. Microorganisms: induction and inhibition of corrosion in metals. *Int Biodeterior Biodegrad*. 2014; 87:66-74.)

iron-oxidizing *Leptothrix discophora* SP-6. They observed that the corrosion of metal exposed to *B. brevis* biofilm was 20-fold lower than the corrosion of metal exposed to non–gramicidin-producing *Paenibacillus polymyxa*.

In spite of successful laboratory experiments on corrosion inhibition by antimicrobial-producing biofilm, the only in situ application of these biofilms did not make the grade. Arps et al.[158] used a consortium of five bacteria (one polymyxin-producing *Bacillus* strain and four gramicidin S–producing *Bacillus* strains) to evaluate microbial corrosion of different metals (304 stainless steel, 1018 carbon steel, and cartridge brass) presented in service water of Three Mile Island Nuclear Power Station. The result indicated that the inhibition of corrosion by the consortium of bacteria was small and pitting of brass was observed in the presence and absence of consortium.

CORROSION INHIBITION WITH BIOLOGICALLY PRODUCED CORROSION INHIBITOR

Bacteria can inhibit corrosion through generating corrosion inhibitor compounds such as polyaspartate and γ-polyglutamate (Fig. 5.9). These two inhibitors bind to positive metal ion through carboxylic groups and form inhibitive metal/peptide complex. Ornek et al.[159] studied the corrosion of aluminum 2024 in presence of two bacterial biofilms: engineered *Bacillus subtilis* biofilm secreting polyaspartate peptides and natural *Bacillus licheniformis* biofilm secreting polyglutamate. Compared with sterile control, pitting was reduced by 90% in presence of *B. licheniformis*, whereas the rate of corrosion reduction by *B. subtilis* biofilm was low.

CORROSION INHIBITION WITH BIOSURFACTANT

Surfactants are commonly used as corrosion inhibitors in engineering systems; however, their high toxicity and nonbiodegradability provide a challenge for many industries where more eco-friendly and biocompatible compounds are needed. Biosurfactants, surfactants produced by microorganisms, which can overcome these limitations, are suitable replacements for their chemical counterparts. Similar to chemical surfactants, biosurfactants can inhibit metal corrosion. Zin et al.[160] demonstrated that rhamnolipid, a biosurfactant produced by *Pseudomonas* sp. PS-17, can passivate D16 T aluminum alloy in a synthetic acid rain solution. Moreover, when the mechanical fracture of passive oxide film occurred, this biosurfactant passivated the metal by creating new juvenile surfaces in the fractured area. Using contact angle measurements and X-ray photoelectron spectroscopy, Shubina et al.[161] observed that the addition of three different concentrations of biosurfactant derived from *Pseudomonas fluorescens* can reduce the corrosion of carbon steel. Adsorption of biosurfactant on the metal surfaces resulted in the formation of a ferric-rich outer layer associated with magnetite. The adsorbed biosurfactant also decreased the hydrophobicity of the carbon surface. Concentration of biosurfactant influences the density and thickness of the biosurfactant conditioning film formed on the metal surface.[162] So while implementing biosurfactant-mediated treatment, the influence of injection dose should be considered.

Attention to the environmental condition is another factor, which should be considered during the biosurfactant appliance. For example, as biosurfactants can promote the formation of gas hydrate, their use as corrosion control agents in fields with high pressure, low temperature, and high gas concentration should be avoided.[163] Dagbert et al.[164] investigated the corrosion behavior of AISI 304 stainless steel in presence of biosurfactant produced by *Pseudomonas fluorescens* (Pf495). This biosurfactant quickly adsorbed on the surface of steel, formed a barrier, and delayed the corrosion of stainless steel. Because biosurfactant adsorption occurred faster than the oxide formation, they hypothesized that this surfactant is a good inhibitor in less aggressive medium (where oxygen is the main oxidizing agent) as compared with aggressive solutions (sodium chloride solution). In the former case, biosurfactant slows down the diffusion of dissolved oxygen, whereas in the latter case, oxide film slows down the diffusion of chloride ion.

Therefore, when an aggressive anion such as chlorine comes close to the metallic surface conditioned by the biosurfactant, the corrosion will enhance quickly owing to lack of protective oxide layer.

Biosurfactants with antiadhesive and antibiofilm activity have a great potential for inhibition of microbiologically influenced corrosion (Fig. 5.10). In 2006, Valle and his coworkers[165] observed that *Escherichia coli* strain CFT073 and all other examined *E. coli* with group II capsule could inhibit the biofilm formation of a wide range of Gram-positive and Gram-negative bacteria by secreting an inhibitory molecule. The inhibitory particle was a capsular polysaccharide with a surface active property, which could be released under undetermined conditions. This compound was inhibiting bacterial adhesion by modifying abiotic surface hydrophobicity/charge and was impairing the cell-surface initial interactions. Additionally, the released biosurfactant was inhibiting biofilm maturation through preventing the bacterial aggregation. Not only by their surface activity but also by antimicrobial activity, some biosurfactants can prevent biofilm formation[166] (Fig. 5.10). For instance, it has been shown that the addition of 1.0% crude biosurfactant of *Bacillus licheniformis* with oil recovery enhancement properties can inhibit the complete growth of SRB bacteria after 3 h.[133]

Remember that not all biosurfactants are having inhibitory properties. Some of them may increase bacterial surface attachment by promoting the bioavailability of nutrient.[134] Thus, while

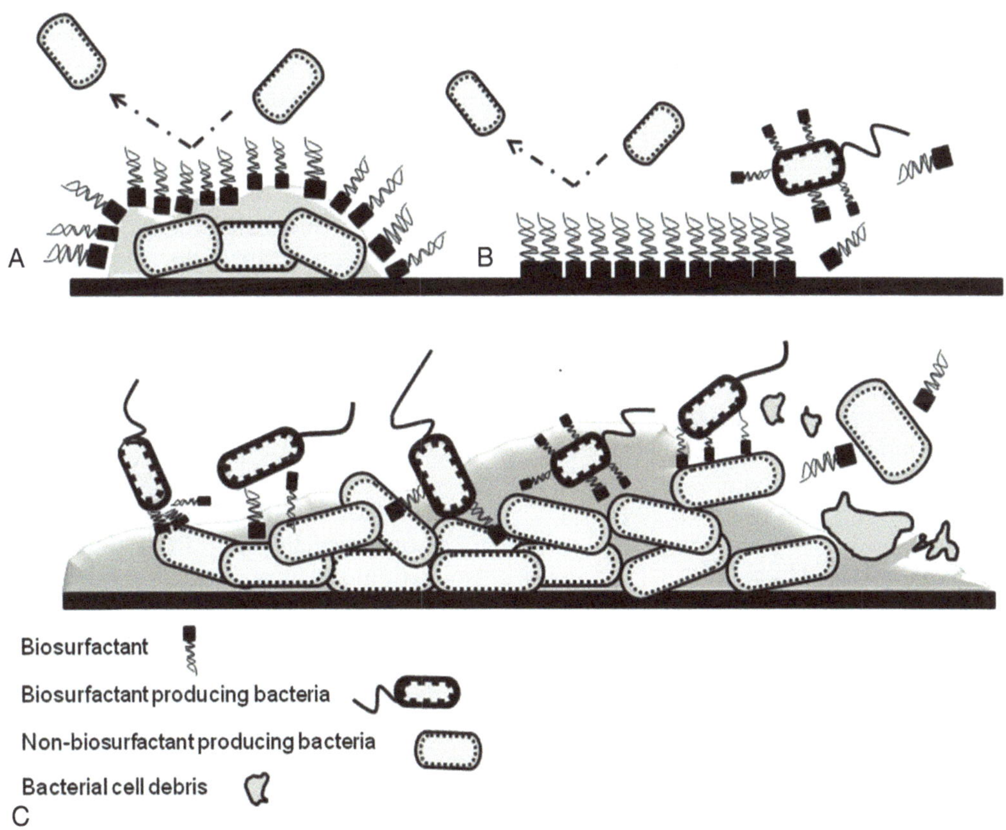

FIG. 5.10 The model determines different mechanisms of corrosion (and biocorrosion) inhibition by biosurfactant. **(A)** Preventing bacterial aggregation and biofilm maturation, **(B)** modification of metal surface and preventing bacterial adhesion, and **(C)** elimination of corrosion-causing bacteria by the antimicrobial activity of biosurfactants.

applying biosurfactants in the oil industry, for example, for the aim of oil recovery, and increase in emulsification, their effect on microbial colonization should be considered. Not only the type of biosurfactant but also the type of bacteria and their physiologic state can affect biosurfactant treatment. Surfactant produced by *Rhodococcus ruber* IEGM231 more effectively prevents the adhesion of actively growing bacteria (in exponential phase) compared with nongrowing (stationary phase) one.[162] Because the physiologic state of bacteria within the biofilm is more close to stationary phase cultures, it is more appropriate to use them for biofilm inhibition and not biofilm treatment.[167]

Although the biosurfactants are already exploited in various industrial areas, widespread incorporation of them is not economically viable for industries. Use of low-cost substrates for biosurfactant production and elimination of the purification process by using biosurfactant crude extract are some approaches, which can make biosurfactant application more industrially practical. Youssef et al.[168] successfully stimulated the in situ production of biosurfactant in an oil reservoir. They injected two lipopeptide biosurfactants producing bacteria, *Bacillus* strain RS-1 and *Bacillus subtilis* subsp. *spizizenii* along with nutrients into two wells of limestone oil reservoir, and obtained high concentration (90 mg/L) of lipopeptide biosurfactants. High yield, low cost, and elimination of purification process make the in situ production of biosurfactant an interesting operation strategy.

CORROSION INHIBITION BY BIOSYNTHESIZED NANOPARTICLES

There are several reports on the antimicrobial activity of a variety of metal and metal oxide nanoparticles such as silver, copper, zinc, and iron.[169–172] Kumar et al.[173] observed that among different sized zero valent iron particles, only nanosized iron inhibited the sulfate reduction in SRB, and granular and microsized iron particles accelerated the rate of sulfate reduction by the SRB. Small size and high surface to volume ratio are two reasons for the stronger antimicrobial activity of NPs than their bulky counterparts. Nanoparticles not only have strong antimicrobial activity but also have shown potential antiadhesive and antibiofilm activity. Liu et al.[174] observed that addition of nanosilver to the superhydrophobic coating reduces the adhesion of bacteria and prevents the formation of bacterial biofilm. Some nanoparticles such as ZnONPs can also inhibit the biofilm formation through disruption of quorum sensing circuits.[175]

Through their antimicrobial, antiadhesive, antibiofilm, and anti-QS activity, nanoparticles can contribute to the inhibition of MIC. It has been demonstrated that the coating of the aluminum surface with an organosiloxane film containing silver nanocrystals resulted in less bacterial surface adhesion, a lower rate of microbiologically influenced corrosion, and higher stability of aluminum.[176] In spite of successful inhibition of biofilm by nanomaterials in the laboratory, application of NPs in natural environment is still a challenging issue for industries because several different factors can affect the interaction between NPs and bacterial biofilm. Some factors such as NPs characteristics, biofilm properties, and environmental factors are influencing the efficiency of the interactions.[177] For instance, engineered NPs are typically functionalized by specific organic ligands for a variety of target applications. However, the presence of natural organic matter in the environment may change the fate and transport of the NPs by forming a coating layer around the NPs. To address these limitations, fabrication of metal surfaces by NPs and delivery of NPs directed by electromagnetic field can be designed. In situ production of nanoparticles is another interesting alternative. Microorganisms such as bacteria, yeast, and fungi can produce NPs.[178–180] Ramanathan et al.[181] showed that there is a strong link between heavy metal resistance and synthesis of nanoparticle in the bacterial system. Microbes contribute to the biosynthesis of nanoparticles not only through heavy metal detoxification but also through energy production.

Various types of NPs (such as Ag, Cu, Pd, U(IV), and Au) have been synthesized by bacteria.[178,182,183] Microbes in both planktonic and sessile forms can produce NPs. Labrenz et al.[184] found that the naturally occurring biofilm of aerotolerant sulfate-reducing bacteria associated with groundwater mediated the formation of 2–5 nm ZnSNPs at a low temperature (8°C) and a pH of 7.2. Although the use of biofilms in the biosynthesis of nanomaterials has not been fully explored, Khalatil et al.[185] showed that the nanomaterials can be synthesized by electrochemically active biofilms (EABs). In their experiment, EABs are initially formed on a carbon paper by dipping it into a mineral salt medium. Anaerobic sludge, which was used as a source of bacterial inoculums, also added to this medium. Then they challenged the EAB with the solution containing $AgNO_3$ as a precursor and sodium acetate as an electron donor in anaerobic condition. They observed that 2 h exposure of the EABs to the mentioned solution resulted in the synthesis of extremely small (1–7 nm) and highly monodispersed spherical AgNPs. Similarly, Khan et al.[186] synthesized cysteine-capped AgNPs by employing the EAB. The biosynthesized particles were having a size range of 5–10 nm, and they effectively inhibited the growth of *Escherichia coli* and *Pseudomonas aeruginosa*.

Compared with the planktonic bacteria, sessile one more effectively forms the NPs. Higher metal resistance, higher biomass, lower precursor concentrations, large surface area, and an inert atmosphere inside the biofilm are some of the properties, which make biofilm favorable for the biosynthesis of NPs.[181,187] Production of NPs inside biofilm minimizes the risk of aggregation, contamination, and oxidization of produced NPs.[188,189] Furthermore, in situ production of NPs by biofilm provides a promising strategy for immobilization of nanoparticles on metal surfaces.

In spite of the significant potential of biofilm for nanoparticle biosynthesis, there are only very few studies on biofilm-mediated synthesis of NPs and no studies on their application particularly for inhibition of corrosion or microbial corrosion.

CORROSION INHIBITION BY HYDROPHOBIC BIOFILM

Coats with liquid-repellent properties have prominent corrosion resistance as well as antifouling, antiicing, and friction-reduction properties.[190–192] Eduok et al.[193] conducted a study to understand the effect of the sol-gel type coating encapsulated with *Bacillus licheniformis* endospores on fouling and corrosion of steel panels. They observed that coat with bacterial endospores has better anticorrosion and antifouling properties compared with coats lacking the same. The enhanced steel resistance was attributed to increasing the coat hydrophobicity because of the presence of bacterial endospore.

Not only the *Bacillus* spore but also their biofilm has hydrophobic properties. The biofilm of *Bacillus subtilis* grown in aerobic condition, displays extreme resistance to liquid wetting and gas penetration[194,195] (Fig. 5.11A). Their biofilms are strong enough to repel up to 80% ethanol, a property that is superior even to Teflon. These remarkable properties of *B. subtilis* biofilm can open new directions in corrosion applications.[195] Although several studies have reported the inhibitory potential of *Bacillus* biofilm, only one that has been done by Guo and his team[197] has correlated the inhibitory properties of biofilm with its hydrophobicity. Guo et al.[197] observed that the *B. subtilis* can increase the corrosion resistance of steel through the formation of a compact, uniform, free of cracks, and hydrophobic biofilm with main components of polysaccharides/TasA amyloid fibers. Hence, with hope to help in and to inspire further discussions, a critical overview about research studies on different aspects of hydrophobic biofilm is given in this section.

A matrix of *B. subtilis* biofilm is composed of an exopolysaccharide and an amyloid fiber-forming protein (tasA), which is assembled with the aid of a small protein called BslA (*Bacillus* surface layer protein A).[198] Epstein et al.[195] reported that exopolysaccharide-deficient (epsH) and matrix protein–deficient (tasA) mutants are lacking the wrinkled architecture of the wild-type biofilms

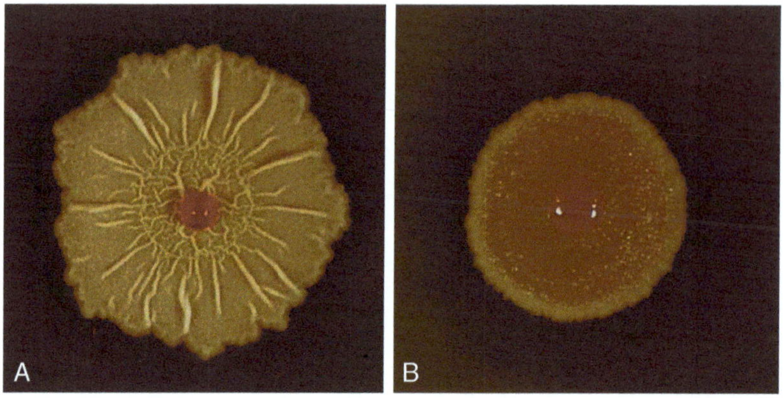

FIG. 5.11 The water repellency of *Bacillus subtilis* biofilm: **(A)** colony of wild-type *B. subtilis* has a hydrophobic surface and **(B)** BslA mutant loses its liquid repellent properties. (Data from Arnaouteli S, MacPhee CE, Stanley-Wall NR, Just in case it rains: building a hydrophobic biofilm the *Bacillus subtilis* way, *Curr Opin Microbiol*. 2016;34:7-12.)

and both of these mutant biofilms were quickly wetted by liquids as compared with the wild-type one. They also reported that although exopolysaccharide and tasA were playing an individual role in biofilm resistance, both of these biomolecules were needed for full repellency. Kobayashi and Iwano[194] observed that the surface of BslA mutant colonies has the wrinkled architecture but the colonies were no longer liquid repellents (Fig. 5.11B). Other studies reported about the important contribution of BslA in the surface hydrophobicity and microroughness of *Bacillus* biofilm.[198,199]

BslA is a structurally defined bacterial hydrophobin (hydrophobins are amphiphilic proteins in filamentous fungi with remarkable properties such as strong adhesion, high surface activity, and self-assembled structure localized at the surface of the complex colonies). This protein has two structural (cap in and cap out) and two functional forms (monomeric and dimeric) by which it attributes in biofilm wettability and non-wettability. Cap in form of BslA presents in the hydrophilic base of biofilm, whereas cap out form of the protein exhibits in the hydrophobic surfaces of biofilm. Monomeric BslA is needed for biofilm structure and dimeric is the form that mediates biofilm surface hydrophobicity. The interactions between BslA molecules also play a role in biofilm hydrophobicity. Unstable lateral interactions are known to be responsible for hydrophobicity if the dimeric protein is present.[200]

It is important to note that the biofilm hydrophobicity is not the stable property and environmental conditions such as electron acceptor availability can alter the biofilm wettability. In oxygen-rich conditions such as surface-exposed region of biofilm, the *B. subtilis* has a hydrophobic biofilm because the cap is out, the BslA predominant form is dimeric, and stable lateral interaction occurs between BslA, whereas in the anoxic conditions similar to the bottom of biofilm, *B. subtilis* loses its hydrophobicity because the protein has its monomeric, cap in form with unstable lateral interactions.[200] In addition to oxygen, the bacterial growth medium also influences biofilm wettability. Biofilm of *B. subtilis* has hydrophilic behavior on LB agar, lotuslike hydrophobic behavior on LBGM agar, and rose petal–like hydrophobic behavior on MSgg agar in the peripheral regions[199] (Fig. 5.12).

BslA is not the only discovered amphiphilic proteins in microbes. Hydrophobins are other amphiphilic proteins responsible for surface hydrophobicity of fungal hyphae and spores, which have surface modification properties.[201] Misra et al.[202] reported that the coating of surface polymer with Sc3 hydrophobin (extracted from wood-rotting fungus *Schizophyllum commune*) resulted in

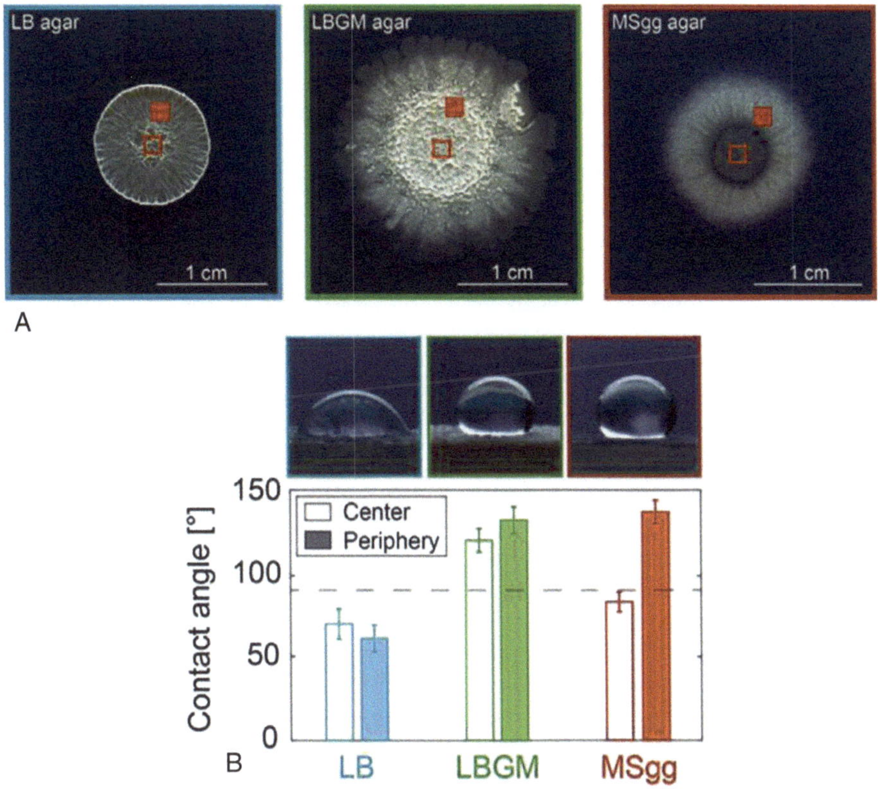

FIG. 5.12 **(A)** The colony of *Bacillus subtilis* NCIB 3610 biofilms grown on different growth media: LB, LBGM, and MSgg. Wetting test has been done on the central and peripheral regions of the bacterial colony, which has been shown by open and closed red square. **(B)** Contact angles of both peripheries and central regions of colony formed on different media. The image of droplet shown in part B is obtained from the peripheral region of each colony. (Data obtained from Werb M, García CF, Bach NC, et al. Surface topology affects wetting behavior of *Bacillus subtilis* biofilms. *NPJ Biofilms and Microbiomes*. 2017;3(1):11.)

a nanoscale reduction of surface friction and also led to increase in surface hydrophobicity of polymer surface. Artini et al.[203] investigated the biofilm formation of three strains of *Staphylococcus epidermidis* (O-47, RP62A, and XX17) on different surfaces (glass, titanium, PTFE, and steel) coated with two fungal hydrophobins (Vmh2 and Pac3). They observed that the coating of material surfaces with both hydrophobins reduced the biofilm formation of three tested strains. In spite of hydrophobin's effect on surface hydrophobicity, Artini and his team believed that the antibiofilm activity was not related to the surface wettability and it was associated with interaction of hydrophobins with eDNA presented in biofilm matrix.

In contrast to the abovementioned study, Rieder et al.[204] reported that coating of the glass surface with recombinantly produced fungal hydrophobin, which led to decrease in surface wettability, did not have any impact on bacterial biofilm in terms of spatial distribution, cell numbers, and population compositions. It has been shown that the type of hydrophobin, bacterial strains, and abiotic surfaces influence the antibiofilm activity of hydrophobin-based coating.

CORROSION INHIBITION BY MICROBIAL ANTIFREEZE PROTEIN

By recalling the Deep Water Horizon disaster in the Gulf of Mexico, which occurred due to hydrate formation, you can perceive the importance of preventing the hydrate formation. Gas hydrates are solid solutions composed of water and appropriate hydrate-forming molecules known as guests.[205] Under conditions of high pressure, low temperature, and high gas concentration, water molecules freeze up to form a solid cagelike network, via hydrogen bonding, around guest molecules such as CH_4, C_2H_6, C_3H_8, and CO_2[205] (Fig. 5.13). Hydrate formation, which can be formed above 0°C, is typically a far greater problem than ice formation in oil and gas pipelines. In addition to reducing/plugging flow in the pipeline, hydrate formation is inducing the pitting and stress cracking corrosions, which are also dangerous to both oil and gas pipelines[206] (Fig. 5.13).

In 2001, Lanoil and his coworkers[207] reported microbial consumption of methane or propane in some of the samples collected from anaerobic hemipelagic mud within the gas hydrate stability zone. Molecular characterization of microbes in one gas hydrate sample named as AT425 showed the presence of several groups of bacteria (Proteobacteria, Actinobacteria, and low-GC Firmicutes) and methanogenic archaea within the gas hydrate. There are reports on association of sulfate-reducing bacteria and methane-oxidizing bacteria with gas hydrate too.[208,209] Arora et al.[163] demonstrated that rhamnolipids biosurfactant produced by *Pseudomonas aeruginosa* strain A11 can reduce the surface tension of water and can promote the formation of methane hydrates. Addition of 1000 ppm biosurfactant to the test system accelerated the methane hydrate formation by 42% and decreased the induction time of hydrate formation by 22.63% as compared with the control system lacking biosurfactant. Furthermore, in presence of rhamnolipid, formation of methane hydrate occurred at higher temperatures. Thus to prevent the gas hydrate formation and subsequent corrosion problem, appliance of surfactant should be minimized in systems potent with hydrate formation.

As bacteria can contribute to the formation of hydrates, they can also attribute in inhibition of hydrate formation. Microorganisms facilitate their survival in low-temperature condition by producing antifreeze proteins (AFPs)[210] (Table 5.3). AFPs inhibit hydrate formation either by causing thermal hysteresis, which is a noncolligative effect defined by a difference between the freezing and melting points of a solution, or by preventing the generation of large ice crystals through boundary migration of smaller ice crystals (or inhibiting the ice recrystallization).[220] Huva et al.[221] studied the effect of five different bacteria (*Chryseobacterium* sp. C14, *Pseudomonas putida, Pseudomonas borealis, Pseudomonas syringae*, and recombinant *Escherichia coli* containing plasmids encoding AFP from the fish *Macrozoarces americanus* and from the plant *Lolium perenne*) with

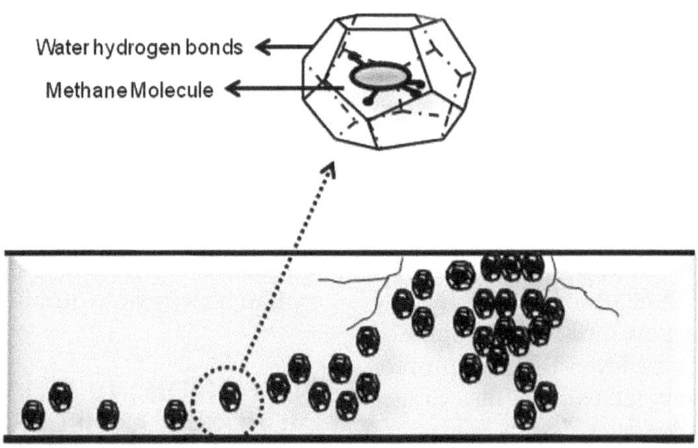

FIG. 5.13 Schematic representation of gas hydrates and their effect on stress cracking corrosions.

TABLE 5.3
List of Antifreeze Proteins Produced by Bacteria

Bacteria	Name of Protein	Reference
Pseudomonas putida GR12-2	Glycoprotein	211
Moraxella sp.	Lipoprotein	212
Pseudomonas fluorescens KUAF-68	Uncharacterized protein	213
Marinomonas primoryensis	*Mp*AFP; Ca^{2+}-dependent protein	214
Colwellia sp. SLW05	*Col*AFP; β-helix protein	215
Flavobacterium xanthum IAM12026	Malate-dependent protein	216
Flavobacteriaceae	Ice-binding protein	217
Flavobacterium frigoris PS1	Ice-binding proteins	218
Noncultured bacteria	EfcIBP; ice-binding protein	219

ice-associating properties on tetrahydrofuran (THF) hydrate crystallization. Among all bacteria, *Chryseobacterium* was the only one which showed the antifreeze activity. *Chryseobacterium* effectively inhibited the THF hydrate crystallization; however, it was unable to prevent the recrystallization of the THF. In another study, Townson et al.[222] evaluated the effect of ice-associating *Chryseobacterium* sp. C14, *Escherichia coli*, and commercial polyvinylpyrrolidone (PVP) on the inhibition of hydrate formation. *E. coli* has shown the greatest inhibitory activity. The average induction time in the presence of *E. coli* was 118.5 h, whereas this time was 38.3 and 30.7 h for *Chryseobacterium* and PVP, respectively. Although *E. coli* did not have any ice-associating activity, the authors hypothesized that *E. coli* could interfere with hydrate formation through secretion of macromolecules and formation of biofilm. There are reports on ice crystallization inhibition ability of bacterial EPS.[223,224] Tamaru et al.[224] observed that EPS plays a crucial role in desiccation and freezing tolerance of a cyanobacterium called *Nostoc commune*. Freezing tolerance of this bacterium was significantly decreased after removal of EPS. Bacterial biofilm acts as a hydrocolloid-like stabilizer and prevents the migration of ice crystals. It can also resist melting and reduce the probability of recrystallization.[225]

INHIBITION OF CORROSION BY TARGETING SIGNALING MOLECULE

To monitor their status in the surrounding environment, bacteria use a variety of intracellular and extracellular signaling molecules. These molecules control most of the bacterial activities including attachment and detachment of bacteria to both biotic and abiotic surfaces. Therefore, one way to control biofilm formation is interrupting their signaling pathways.

Corrosion Inhibition by Quorum Sensing Inhibitors

Many aspects of biofilm formation and maturation such as initial bacterial colonization, EPS production, biofilm homogeneity, and biofilm thickness are regulated through quorum sensing (QS) process.[225–228] Therefore, inhibition of QS and its impact on biofilm inhibition could offer great potential for mitigation of MIC.

QS is getting interrupted by several different strategies such as inhibition of signal synthesis, degradation of signals, inhibition of signal-receptor binding, and inhibition of binding of the signal transduction pathway[229] (Fig. 5.14). Theoretically, blockage of signal synthesis is one of the simplest methods for inhibition of QS pathways; however, in practice, only very few compounds with the ability to inhibit the signal

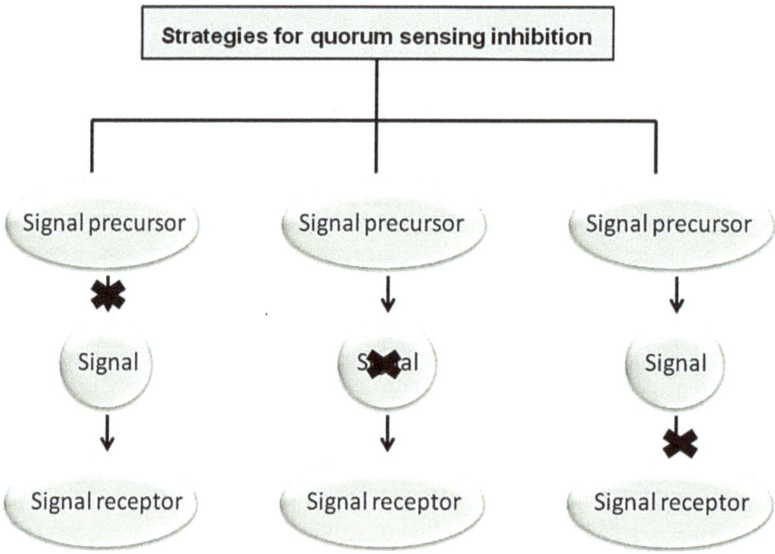

FIG. 5.14 Different strategies for inhibition of quorum sensing process.

synthesis have been identified. Triggering the binding of a signal molecule to its receptor is one of the major strategies for QS inhibition. In this approach, compounds that have structural similarity with autoinducers are used as QS inhibitors (QSIs). These inhibitors that block the receptors can be of either natural or synthetic origin.[230,231] Phenylethylamine, a secondary metabolite produced by *Halobacillus salinus* bacterium, is competing with AHL for the receptor binding site.[232] Similarly, a lactone-containing compound extracted from *Streptomyces* sp. was showing signal-receptor binding inhibitory activity.[233] Ibacache–Quiroga et al.[234] demonstrated that a hydrocarbon-degrading bacterium, *Bacterium cobetia* mm1ida2h-1, can produce a biosurfactant with the ability to interfere with the receptor-binding process. Direct destruction of a signaling molecule is another important approach, which can hamper bacterial communication. Quorum quenching (QQ) enzymes such as AHL lactonase, AHL acylase, deaminase, and decarboxylase can be used for complete degradation or inactivation of AHL signals.[229] AHL lactonase and decarboxylase degrade the AHL molecule by hydrolyzing lactone ring, whereas AHL acylase and deaminase cleave the acyl side chain of AHL molecule.[235] Lin et al.[236] demonstrated that the expression of the recombinant gene responsible for the synthesis of AHL acylase in *Pseudomonas aeroginosa* PAO1 quenched some QS-related activities of PAO1. Table 5.4 summarizes the list of quorum quenching enzymes produced by bacteria.

Although inhibition of MIC by inhibiting QS is a green strategy with various advantages such as the low risk of resistance development in bacteria and high antibiofilm efficiency, MIC treatment with this method is not without limitations.[231,238,239] Many QSIs have narrow spectrum (i.e., specific inhibitor is needed for targeting particular bacteria).[240] For instance, AHL acylase from *Streptomyces* sp. can only degrade AHLs containing acyl chains with six or more carbons.[241] AHL acylase produced by *Pseudomonas aeruginosa* can specifically degrade 3-oxoC12-HSL, not C4-HSL.[242] QSIs dissociate the biofilm architecture without the significant killing of the bacterial cell.[229] The release of live bacteria into environment reduces the efficacy of MIC treatment, as they can reestablish the biofilm. However, this risk can be reduced by a combination of QSIs with a biocide. In this case, treatment efficacy will improve because QSI will increase the susceptibility of bacteria to antimicrobial compounds.

In spite of all the disadvantages, laboratory studies have shown successful biofilm treatment by this method. By using acylase enzyme, which

TABLE 5.4
Bacterial Enzymes as Quorum Sensing Inhibitors

Source of Quorum Sensing Inhibitor	Enzyme	Degraded Quorum Sensing Signal
Bacillus sp. strain 240B1	Lactonase	AHLs
Bacillus thuringiensis	Lactonase	AHLs
Stappia[a] sp. strains 5, 176, and 97-1	Lactonase	AHLs
Oceanobacillus strains 30, 172, and 97	Lactonase	AHLs
Halomonas sp. strain 33	Lactonase	AHLs
Tenacibaculum discolor strain 20J	Acylase/lactonase	AHLs
Hyphamonas sp. DG895	Acylase/lactonase	C4HSL and 3OC12
Alteromonas sp. strain 168	Acylase	C4HSL and 3OC12
Bacillus megaterium	AHL oxidase	C4HSL and 3OC12
Bacillus circulans strain 24	Different from lactonase[b]	C4HSL and 3OC12HSL
Bacillus pumilus S8-07	AHL acylase	3OC12HSL
Ralstonia sp. XJ12B	AHL acylase	Long-chain AHLs
Pseudomonas aeruginosa PAO1	AHL acylase	Long-chain AHLs
Rhodococcus erythropolis strain W2	AHL lactonase	AHLs
	Acylase (amidohydrolase)	AHLs
	Oxidoreductase activity	3-oxo-N-AHLs
Burkholderia strain GG4	AHL oxidoreductase	3OC6HSL
Agrobacterium tumefaciens	AHL lactonase	AHLs
Arthrobacter sp. IBN110	AHL lactonase	AHLs
Acinetobacter sp. strain C1010	Lactonase	AHLs

[a] Formerly belonging to *Agrobacterium*.
[b] AHLs could not be recovered by acidification, may be similar to AHL oxidase of *B. megaterium*.
From Kalia VC. Quorum sensing inhibitors: an overview. *Biotechnol Adv*. 2013;31(2):224-245.

was targeting AHL molecules, Kim et al.[243] could control the biofilm formation of *Pseudomonas putida*, which was isolated from a fouled RO membrane. QSIs are working in a concentration-dependent manner so diffusion of QSI in open systems would reduce the treatment efficacy.[244] Kim et al.[245] offered an alternative for preventing the diffusion of inhibitors. They immobilized a quorum quenching enzyme onto a nanofiltration membrane to mitigate biofouling in a nanofiltration process. The immobilized enzyme inhibited the biofilm formation (by affecting on EPS production) while more than 90% of its activity was maintained for more than 20 iterative cycles of reaction and washing procedure. By immobilizing the acylase I (QQ enzyme) on magnetic particles, Yeon et al.[246] prepared a magnetic enzyme carrier having great potential in inhibition of membrane biofouling. High extraction and purification cost is another drawback for MIC treatment by QS inhibitors. This problem can be solved by using the quorum quenching bacteria. Oh et al.[247] studied the effect of recombinant *Escherichia coli* with quorum quenching ability on the membrane biofouling caused by *Pantoea stewartii*. They introduced the QQ bacteria into reverse osmosis system by two strategies (direct injection and immobilization within a microfilter) and observed that, in both conditions, QQ bacteria could control the membrane biofouling by reducing the biofilm formation of *P. stewartii*. Kim et al.[248] observed that free-moving alginate beads entrapped with indigenous quorum quenching *Rhodococcus* sp. BH4 can control the membrane biofouling by inhibiting the biofilm formation. Quorum quenching and physical washing of biofilm through the collision of beads with the membrane was responsible for inhibition of biofilm (Fig. 5.15).

Despite the importance of QS and QS inhibition, this mechanism is still unknown in many corrosion-causing bacteria including sulfate-reducing bacteria, iron-reducing bacteria, and sulfur-oxidizing bacteria. Therefore for successful treatment of MIC with QSI, adequate attention should be devoted toward the identification of this system in bacteria affecting the deterioration of metals.

Inhibition of Corrosion by Targeting c-di-GMP

c-di-GMP is a secondary messenger that coordinates the "lifestyle transition" from free-floating bacteria to sessile form and vice versa. Bacterial biofilm contains a high concentration of c-di-GMP, whereas planktonic bacteria have a low level of c-di-GMP[249] (Fig. 5.16). The concentration of c-di-GMP in a cell is under the control of two enzymes. c-di-GMP is synthesized by diguanylate cyclase (DGC) enzymes and degraded by phosphodiesterase (PDE) enzymes.[250] These enzymes contain various N-terminal sensory domains to respond to environmental signals such as nitric oxide.[251] It has been shown that the nitric oxide triggers the bacterial biofilm even at low, nontoxic concentration.[252] Barraud et al.[253] demonstrated that biofilm dispersal of *Pseudomonas aeruginosa* in response to nitric oxide is associated with the c-di-GMP signaling. Nitric oxide stimulates the PDE activity, which subsequently reduces the intercellular level of c-di-GMP in cell and induces the dispersal of bacterial biofilm.

As nitric oxide is an obligatory bacterial nitrification intermediate, triggering the c-di-GMP pathways by nitric oxide–producing bacteria could be an attractive target for elimination of old biofilm.[254] Unfortunately, ever since, no study has investigated the effect of NRB bacteria on biofilm dispersal with regard to the c-di-GMP pathway.

CORROSION INHIBITION BY MICROBIAL FOOTPRINT

For the first time, Marshal et al.[255] observed that when *Pseudomonas* detaches from a glass surface, some polymeric substances will remain on the

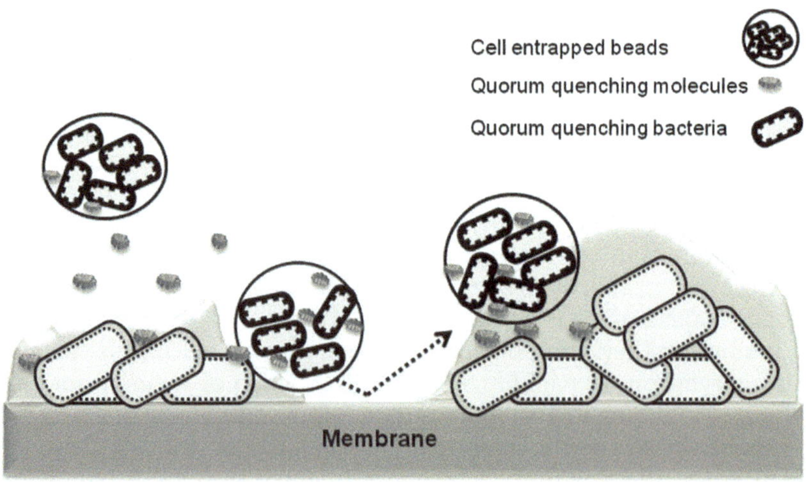

FIG. 5.15 Schematic representation of biofilm inhibition by quorum quenching bacteria entrapped in alginate beads. Cell-entrapped beads are inhibiting biofilm by two mechanisms: (1) secretion of quorum quenching molecules and (2) physical removal of biofilm through collision with biofilm.

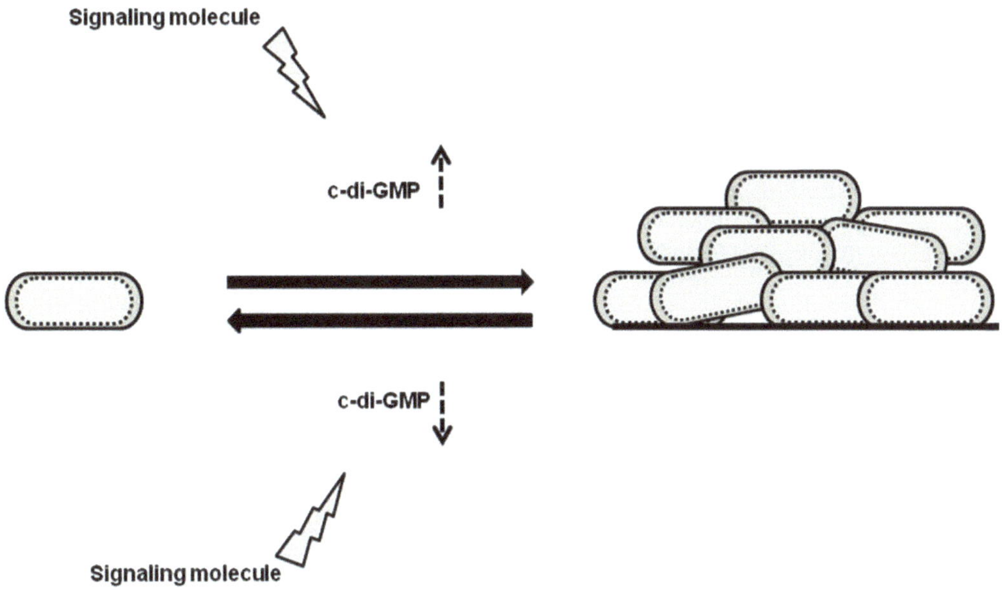

FIG. 5.16 Schematic representation of the effect of c-di-GMP on biofilm formation and biofilm dispersal.

glass surface. These adhesive macromolecules left on substrate surfaces after removal of the attached cells are called "microbial footprints." The microbial footprint will remain after microbial detachment or mechanical removal of microbes (shear force and sonication)[255–257] (Fig. 5.17). Presence of microbial footprint on the inert surface has been reported even after enzymatic treatment of biofilm.[258]

Microbes produce these adhesive macromolecules during initial attachment because these multipurpose bridging polymers enable bacteria to overcome the repulsive properties of surfaces and make a strong attachment with the surfaces.[259] In addition, it seems during the detachment, bacteria use their footprint to label the relevant surfaces and send the message "already exploited" to other bacteria.[256,260] Therefore, microbial footprint or "microbial conditioning film," which is mainly composed of bacterial EPS, can be used for protection of metal surface against the settlement of bacteria (Fig. 5.17). In 2002, Gomez-Suarez et al.[261] investigated adhesion of two *Pseudomonas aeruginosa* strains; SG81 an EPS-producing and SG81R1 a non–EPS-producing *P. aeruginosa* strains, on substrate precoated with isolated EPS of strain SG81. They observed that precoating of metal with EPS has a negative impact on adhesion of both strains. Later, they studied the effect of a footprint left by these strains on their readhesion. They allowed bacteria to form a biofilm on the glass surface, mechanically detached their biofilm, and then allowed bacteria to redeposit on the surface. Adhesion of both bacteria has reduced, and the reduction rate was higher for EPS-producing strain because this strain left the higher amount of EPS on the glass surface compared with non–EPS-producing one. Although the reattachment of nonproducing strain was higher compared with the EPS-producing strain, this rate becomes similar to the readhesion of EPS-producing strain after six cycles of detachment-deposition. Stadler et al.[262] isolated and purified EPS from biofilm of sulfate-reducing bacteria and then used it for coating the surface of stainless steel at the different temperature. Later, the steel surface was exposed to the SRB-containing solution. They observed that, although the EPS was mounted as a discontinuous layer on the steel surface, this layer reduced the settlement of SRB on the metal surface. Hwang et al.[263] also observed that regardless of the bacterial strain, the precoating of slide surface by EPS extracted from *Burkholderia cepacia* nd

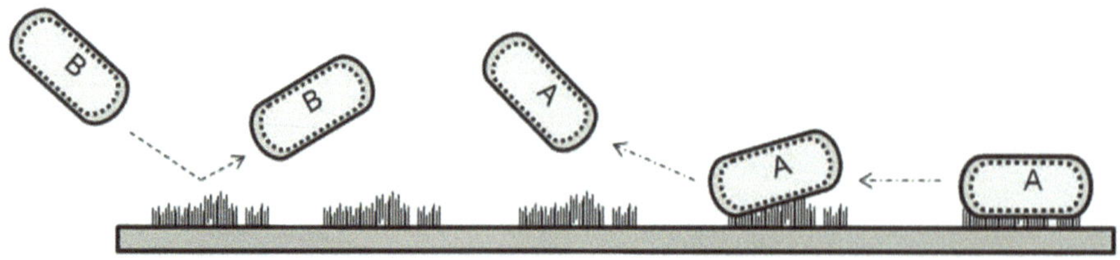

FIG. 5.17 **(A)** Microbial footprint: adhesive material remains on the surface after microbial detachment; **(B)** microbial footprint prevents adhesion of bacteria.

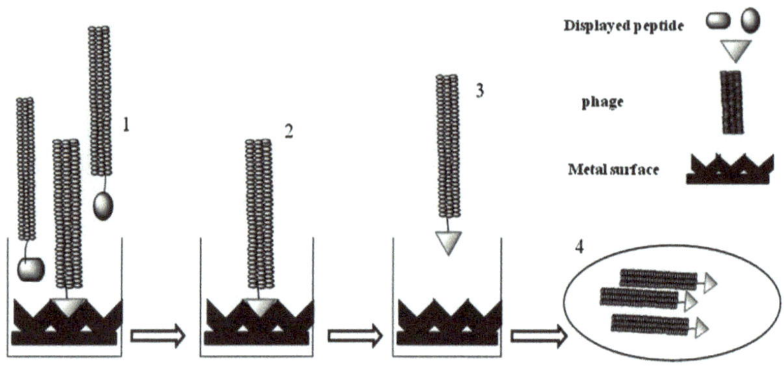

FIG. 5.18 Phage peptide displays technology for coating the metal surface; step 1: binding of phages displaying desired peptides to the metal surface; step 2: removal of unbound phage by washing; step 3: elution of the bounded phage; step 4: amplification of eluted phage in the bacterial cell. (Data from Alasvand Zarasvand K, Rai VR. Microorganisms: induction and inhibition of corrosion in metals. *Int Biodeterior Biodegrad*. 2014;87:66–74.)

Pseudomonas aeruginosa reduced the bacterial adhesion as compared with bare slides. Bacterial adhesion to *P. aeruginosa* EPS–coated slide was lower than the adhesion to *B. cepacia* EPS because of higher protein content in *P. aeruginosa* EPS, which was resulting in higher repulsion.

CORROSION INHIBITION BY MICROBIAL CELL SURFACE DISPLAY

It has been shown that a peptide isolated from a type IV pilin receptor binding domain *of Pseudomonas aeruginosa* has an excellent anticorrosion effect.[264] Amino acids have corrosion inhibitors' property. Hluchan et al.[265] investigated the inhibitory effect of 22 amino acids on iron corrosion in 1 M hydrochloric acid and observed the variable inhibitive behavior of amino acids. Some amino acids such as cystine and cysteine accelerated corrosion, whereas some amino acids such as tryptophan reduced corrosion. Amino acids with a longer hydrocarbon chain and additional amino groups showed higher inhibition, whereas shorter-chain amino acids such as aspartic acid exhibited only a 50% reduction in corrosion rate. In another study, Muller[266] observed that although aspartic acid is inhibiting aluminum and not zinc, polyaspartic acid is not protecting aluminum, but it inhibits corrosion of zinc.

Microbial cell surface display is a technique that facilitates the screening of peptide. In this technique, the desired peptide is expressing on the surface of microbe (phage, bacteria, or yeast). Later, the microbe that is expressing the peptide will bind to substrates for which the expressed peptide has an affinity.[267] The cell surface display technique can be used for screening of peptide, which has metal corrosion inhibitory activity (Fig. 5.18). Samuelson et al.[268] generated recombinant *Staphylococcus xylosus* and *Staphylococcus carnosus* strains, which were expressing chimeric proteins containing polyhistidyl peptides on their surface. Expression of this surface protein gave metal binding capacity to these bacteria. To find a novel peptide inhibitor for protecting aluminum and mild steel against corrosion, Zuo et al.[269] used a random phage display library composed of 12 amino acid peptides. They observed that the peptides that are rich in amino acids containing hydroxyl residues such as serine, threonine, and histidine have better interaction with the metal surface.

REFERENCES

1. Martin MO. Predatory prokaryotes: an emerging research opportunity. *J Mol Microbiol Biotechnol.* 2002;4(5):467–477.
2. Jurkevitch E. Predatory behaviors in bacteria — diversity and transitions. *Microbe Mag.* 2007;2(2): 67–73.
3. Richards G, Fay J, Uknalis J, Olanya OM, Watson MA. Purification and host specificity of predatory *Halobacteriovorax* isolated from seawater. *Appl Environ Microbiol.* 2016;82(3):922–927.
4. Oyedara OO, Luna-Santillana EDJD, Olguin-Rodriguez O, et al. Isolation of *Bdellovibrio* sp. from soil samples in Mexico and their potential applications in control of pathogens. *Microbiologyopen.* 2016;5(6):992–1002.
5. Li N, Williams HN. 454 Pyrosequencing reveals diversity of *Bdellovibrio* and like organisms in fresh and salt water. *Ant Van Leeuwenhoek.* 2015; 107(1):305–311.
6. Williams HN, Piñeiro S. Ecology of the predatory *Bdellovibrio* and like organisms. In: *Predatory Prokaryotes.* Springer Berlin Heidelberg; 2006: 213–248.
7. Ferguson M, Núñez M, Kim HJ, et al. Spatially organized films from *Bdellovibrio bacteriovorus* prey lysates. *Appl Environ Microbiol.* 2014;80(23): 7405–7414.
8. Rendulic S, Jagtap P, Rosinus A, et al. A predator unmasked: life cycle of *Bdellovibrio bacteriovorus* from a genomic perspective. *Science.* 2004; 303(5658):686–692.
9. Sockett RE, Lambert C. *Bdellovibrio* as therapeutic agents: a predatory renaissance? *Nat Rev Microbiol.* 2004;2(8):669–675.
10. Varon M, Seijffers J. Symbiosis-independent and symbiosis-incompetent mutants of *Bdellovibrio bacteriovorus* 109J. *J Bacteriol.* 1975;124(3): 1191–1197.
11. Pantanella F, Iebba V, Mura F, et al. Behaviour of *Bdellovibrio bacteriovorus* in the presence of gram-positive *Staphylococcus aureus. New Microbiol.* 2018;41(2):145–152.
12. Iebba V, Totino V, Santangelo F, et al. *Bdellovibrio bacteriovorus* directly attacks *Pseudomonas aeruginosa* and *Staphylococcus aureus* cystic fibrosis isolates. *Front Microbiol.* 2014;5:280.
13. Monnappa AK, Dwidar M, Seo JK, Hur J-H, Mitchell RJ. *Bdellovibrio bacteriovorus* inhibits *Staphylococcus aureus* biofilm formation and invasion into human epithelial cells. *Sci Rep.* 2015;4(1): 3811.
14. Schoeffield A, Williams H, Turng B-F, Falkler W. A comparison of the survival of intraperiplasmic and attack phase *bdellovibrios* with reduced oxygen. *Microb Ecol.* 1996;32(1):35–46.
15. Kadouri D, O'Toole GA. Susceptibility of biofilms to *Bdellovibrio bacteriovorus* attack. *Appl Environ Microbiol.* 2005;71(7):4044–4051.
16. Qiu L, Mao Y, Gong A, Zhang W, Cao Y, Tong L. Inhibition effect of *Bdellovibrio bacteriovorus* on the corrosion of X70 pipeline steel induced by sulfate-reducing bacteria. *Anti-corrosion Methods Mater.* 2016;63(4):269–274.
17. Markelova NY. Predacious bacteria, *Bdellovibrio* with potential for biocontrol. *Int J Hyg Environ Health.* 2010;213(6):428–431.
18. Koval SF, Bayer ME. Bacterial capsules: no barrier against *Bdellovibrio. Microbiology.* 1997;143(3): 749–753.
19. Van Essche M, Quirynen M, Sliepen I, et al. Killing of anaerobic pathogens by predatory bacteria. *Mol Oral Microbiol.* 2011;26(1):52–61.
20. Shemesh Y, Jurkevitch E. Plastic phenotypic resistance to predation by *Bdellovibrio* and like organisms in bacterial prey. *Environ Microbiol.* 2003; 6(1):12–18.
21. Koval SF, Hynes SH, Flannagan RS, Pasternak Z, Davidov Y, Jurkevitch E. *Bdellovibrio exovorus* sp. nov., a novel predator of *Caulobacter crescentus. Int J Syst Evol Microbiol.* 2013;63(Pt 1): 146–151.
22. Müller S, Strack SN, Hoefler BC, Straight PD, Kearns DB, Kirby JR. Bacillaene and sporulation protect *Bacillus subtilis* from predation by *Myxococcus xanthus. Appl Environ Microbiol.* 2014;80(18): 5603–5610.
23. Müller S, Strack SN, Ryan SE, Kearns DB, Kirby JR. Predation by *Myxococcus xanthus* induces *Bacillus subtilis* to form spore-filled megastructures. *Appl Environ Microbiol.* 2015;81(1):203–210.

24. Massana R, Jürgens K. Composition and population dynamics of planktonic bacteria and bacterivorous flagellates in seawater chemostat cultures. *Aquat Microb Ecol*. 2003;32(1):11–22.
25. Harcombe WR, Bull JJ. Impact of phages on two-species bacterial communities. *Appl Environ Microbiol*. 2005;71(9):5254–5259.
26. Varon M, Zeigler B. Bacterial predator-prey interaction at low prey density. *Am Soc Microbiol*. 1978;36(1):11–17.
27. Kim E-H, Dwidar M, Mitchell RJ, Kwon Y-N. Assessing the effects of bacterial predation on membrane biofouling. *Water Res*. 2013;47(16): 6024–6032.
28. Seccareccia I, Kovács ÁT, Gallegos-Monterrosa R, Nett M. Unraveling the predator-prey relationship of *Cupriavidus necator* and *Bacillus subtilis*. *Microbiol Res*. 2016;192:231–238.
29. Muñoz-Dorado J, Marcos-Torres FJ, García-Bravo E, Moraleda-Muñoz A, Pérez J. Myxobacteria: moving, killing, feeding, and surviving together. *Front Microbiol*. 2016;7:781.
30. Seccareccia I, Kost C, Nett M. Quantitative analysis of Lysobacter predation. *Appl Environ Microbiol*. 2015;81(20):7098–7105.
31. Shilo M. Lysis of blue-green algae by myxobacter. *J Bacteriol*. 1970;104(1):453–461.
32. Ashton PJ, Robarts RD. Apparent predation of *microcystis aeruginosa* kütz. emend elenkin by a *saprospira* -like bacterium in a hypertrophic lake (Hartbeespoort dam, South Africa). *J Limnol Soc South Afr*. 1987;13(1):44–47.
33. Banning EC, Casciotti KL, Kujawinski EB. Novel strains isolated from a coastal aquifer suggest a predatory role for flavobacteria. *Fems Microbiol Ecol*. 2010;73(2):254–270.
34. Quinn GR, Skerman VBD. Herpetosiphon—nature's scavenger? *Curr Microbiol*. 1980;4(1): 57–62.
35. Guerrero R, Pedros-Alio C, Esteve I, Mas J, Chase D, Margulis L. Predatory prokaryotes: predation and primary consumption evolved in bacteria. *Proc Natl Acad Sci USA*. 1986;83(7): 2138–2142.
36. Shatzkes K, Tang C, Singleton E, et al. Effect of predatory bacteria on the gut bacterial microbiota in rats. *Sci Rep*. 2017;7:43483.
37. Makkar NS, Casipa LE. *A Nonobligate Bacterial Predator of Bacteria in Soil?* 1987.
38. Casida Jr LE. Interaction of Agromyces ramosus with other bacteria in soil. *Appl Environ Microbiol*. 1983;46(4):881–888.
39. Casida LE. Bacterial predators of *Micrococcus luteus* in soil. *Appl Environ Microbiol*. 1980;39(5): 1035–1041.
40. Casida LE. Minireview: *Nonobligate* bacterial predation of bacteria in soil. *Microb Ecol*. 1988; 15(1):1–8.
41. Kumbhar C, Mudliar P, Bhatia L, Kshirsagar A, Watve M. Widespread predatory abilities in the genus streptomyces. *Arch Microbiol*. 2014;196(4): 235–248.
42. K.K. Rashidan and D.F. Bird, Role of predatory bacteria in the termination of a cyanobacterial bloom. *Microb Ecol* 2001;41(3):97–105
43. Enos BG, Anthony MK, DeGiorgis JA, Williams LE. Prey range and genome evolution of Halobacteriovorax marinus predatory bacteria from an Estuary. *mSphere*. 2018;3(1).
44. Pineiro SA, Williams HN, Stine OC. Phylogenetic relationships amongst the saltwater members of the genus Bacteriovorax using rpoB sequences and reclassification of Bacteriovorax stolpii as Bacteriolyticum stolpii gen. nov., comb. nov. *Int J Syst Evol Microbiol*. 2008;58(5):1203–1209.
45. Jurkevitch E. *Predatory Prokaryotes : Biology, Ecology and Evolution*. Springer; 2007:251.
46. Hol FJH, Rotem O, Jurkevitch E, Dekker C, Koster DA. Bacterial predator-prey dynamics in microscale patchy landscapes. *Proc Biol Sci*. 2016; 283(1824):pii:20152154.
47. Ranu RS, Gallegos R, Althauser M, Wolfe L. The structural proteins of *Bdellovibrio bacteriovorus* bacteriophage MAC-1. *World J Eng Technol*. 2016; 04(03):7–13.
48. Kadouri DE, Tran A. Measurement of predation and biofilm formation under different ambient oxygen conditions using a simple gasbag-based system. *Appl Environ Microbiol*. 9 2013;79(17): 5264–5271.
49. Markelova N. Effect of toxic pollutants on *Bdellovibrio*. *Process Biochem*. 2002;37(10): 1177–1181.
50. Nicastro J, Wong S, Khazaei Z, Lam P, Blay J. *Bacteriophage Applications-Historical Perspective and Future Potential*. Springer; 2016.
51. Eydal HSC, Jägevall S, Hermansson M, Pedersen K. Bacteriophage lytic to *Desulfovibrio aespoeensis* isolated from deep groundwater. *ISME J*. 2009;3(10):1139–1147.
52. Sillankorva S, Neubauer P, Azeredo J. Phage control of dual species biofilms of *Pseudomonas fluorescens* and *Staphylococcus lentus*. *Biofouling*. 2010;26(5):567–575.

53. Motlagh AM, Bhattacharjee AS, Goel R. Biofilm control with natural and genetically-modified phages. *World J Microbiol Biotechnol*. 2016;32(4):67.
54. Marzari R, Sblattero D, Righi M, Bradbury A. Extending filamentous phage host range by the grafting of a heterologous receptor binding domain. *Gene*. 1997;185(1):27–33.
55. Lu TK, Collins JJ. Dispersing biofilms with engineered enzymatic bacteriophage. *Proc Natl Acad Sci*. 2007;104(27):11197–11202.
56. Scholl D, Adhya S, Merril C. *Escherichia coli* K1's capsule is a barrier to bacteriophage T7. *Appl Environ Microbiol*. 2005;71(8):4872–4874.
57. Sass P, Bierbaum G. Lytic activity of recombinant bacteriophage phi11 and phi12 endolysins on whole cells and biofilms of *Staphylococcus aureus*. *Appl Environ Microbiol*. 2007;73(1):347–352.
58. Rodríguez L, Martínez B, Zhou Y, Rodríguez A, Donovan DM, García P. Lytic activity of the virion-associated peptidoglycan hydrolase HydH5 of *Staphylococcus aureus* bacteriophage vB_SauS-phiIPLA88. *BMC Microbiol*. 2011;11:138.
59. Pei R, Lamas-Samanamud GR. Inhibition of biofilm formation by T7 bacteriophages producing quorum-quenching enzymes. *Appl Environ Microbiol*. 2014;80(17):5340–5348.
60. Westwater C, Kasman LM, Schofield DA, et al. Use of genetically engineered phage to deliver antimicrobial agents to bacteria: an alternative therapy for treatment of bacterial infections. *Antimicrob Agents Chemother*. 2003;47(4):1301–1307.
61. Yacoby I, Shamis M, Bar H, Shabat D, Benhar I. Targeting antibacterial agents by using drug-carrying filamentous bacteriophages. *Antimicrob Agents Chemother*. 2006;50(6):2087–2097.
62. Yoichi M, Abe M, Miyanaga K, Unno H, Tanji Y. Alteration of tail fiber protein gp38 enables T2 phage to infect *Escherichia coli* O157:H7. *J Biotechnol*. 2005;115(1):101–107.
63. Lu TK, Collins JJ. Engineered bacteriophage targeting gene networks as adjuvants for antibiotic therapy. *Proc Natl Acad Sci USA*. 2009;106(12):4629–4634.
64. M. N, Rapson ME, Burden FA, Glancy LP, Hodgson DA. *Bacteriophages Useful for Therapy and Prophylaxis of Bacterial Infections*. 2003.
65. Hagens S, Blasi U. Genetically modified filamentous phage as bactericidal agents: a pilot study. *Lett Appl Microbiol*. 2003;37(4):318–323.
66. Ganegama Arachchi GJ, Cridge AG, Dias-Wanigasekera BM, et al. Effectiveness of phages in the decontamination of *Listeria monocytogenes* adhered to clean stainless steel, stainless steel coated with fish protein, and as a biofilm. *J Ind Microbiol Biotechnol*. 2013;40(10):1105–1116.
67. Fu W, Forster T, Mayer O, Curtin JJ, Lehman SM, Donlan RM. Bacteriophage cocktail for the prevention of biofilm formation by *Pseudomonas aeruginosa* on catheters in an in vitro model system. *Antimicrob Agents Chemother*. 2010;54(1):397–404.
68. Hyman P, Abedon ST. Bacteriophage host range and bacterial resistance. *Adv Appl Microbiol*. 2010;70:217–248.
69. Sutherland IW, Hughes KA, Skillman LC, Tait K. The interaction of phage and biofilms. *Fems Microbiol Lett*. 2004;232(1):1–6.
70. Meader E, Mayer MJ, Steverding D, Carding SR, Narbad A. Evaluation of bacteriophage therapy to control *Clostridium difficile* and toxin production in an in vitro human colon model system. *Anaerobe*. 2013;22:25–30.
71. Gödeke J, Paul K, Lassak J, Thormann KM. Phage-induced lysis enhances biofilm formation in Shewanella oneidensis MR-1. *ISME J*. 2011;5(4):613–626.
72. Nale JY, Spencer J, Hargreaves KR, et al. Bacteriophage combinations significantly reduce *Clostridium difficile* growth in vitro and proliferation in vivo. *Antimicrob Agents Chemother*. 2016;60(2):968–981.
73. Labrie SJ, Samson JE, Moineau S. Bacteriophage resistance mechanisms. *Nat Rev Microbiol*. 2010;8(5):317–327.
74. Briandet R, Lacroix-Gueu P, Renault M, et al. Fluorescence correlation spectroscopy to study diffusion and reaction of bacteriophages inside biofilms. *Appl Environ Microbiol*. 2008;74(7):2135–2143.
75. Stummeyer K, Schwarzer D, Claus H, Vogel U, Gerardy-Schahn R, Muhlenhoff M. Evolution of bacteriophages infecting encapsulated bacteria: lessons from *Escherichia coli* K1-specific phages. *Mol Microbiol*. 2006;60(5):1123–1135.
76. Schmerer M, Molineux IJ, Bull JJ. Synergy as a rationale for phage therapy using phage cocktails. *PeerJ*. 2014;2:e590.

77. Son J-S, Lee S-J, Jun SY, et al. Antibacterial and biofilm removal activity of a podoviridae *Staphylococcus aureus* bacteriophage SAP-2 and a derived recombinant cell-wall-degrading enzyme. *Appl Microbiol Biotechnol*. 2010;86(5):1439–1449.
78. Harper D, Parracho H, Walker J, et al. Bacteriophages and biofilms. *Antibiotics*. 2014;3(3): 270–284.
79. Gutiérrez D, Ruas-Madiedo P, Martínez B, Rodríguez A, García P. Effective removal of *Staphylococcal* biofilms by the endolysin LysH5. *PLoS One*. 2014;9(9):e107307.
80. Pearl S, Gabay C, Kishony R, Oppenheim A, Balaban NQ. Nongenetic individuality in the host–phage interaction. *PLoS Biol*. 2008;6(5): e120.
81. Lehman SM, Donlan RM. Bacteriophage-mediated control of a two-species biofilm formed by microorganisms causing catheter-associated urinary tract infections in an in vitro urinary catheter model. *Antimicrob Agents Chemother*. 2015; 59(2):1127–1137.
82. Hadas H, Einav M, Fishov I, Zaritsky A. Bacteriophage T4 development depends on the physiology of its host *Escherichia coli*. *Microbiology*. 1997; 143(1):179–185.
83. Rodríguez-Rubio L, Martínez B, Donovan DM, Rodríguez A, García P. Bacteriophage virion-associated peptidoglycan hydrolases: potential new enzybiotics. *Crit Rev Microbiol*. 2013;39(4): 427–434.
84. Fischetti VA. Bacteriophage lysins as effective antibacterials. *Curr Opin Microbiol*. 2008;11(5): 393–400.
85. Nobrega FL, Costa AR, Kluskens LD, Azeredo J. Revisiting phage therapy: new applications for old resources. *Trends Microbiol*. 2015;23(4): 185–191.
86. Shen Y, Köller T, Kreikemeyer B, Nelson DC. Rapid degradation of *Streptococcus pyogenes* biofilms by PlyC, a bacteriophage-encoded endolysin. *J Antimicrob Chemother*. 2013;68(8): 1818–1824.
87. Mojica KD, Brussaard CP. Factors affecting virus dynamics and microbial host - virus interactions in marine environments. *Fems Microbiol Ecol*. 2014;89(3):495–515.
88. Kiefer B, Dahl JL. Disruption of *Mycobacterium smegmatis*; biofilms using bacteriophages alone or in combination with mechanical stress. *Adv Microbiol*. 2015;05(10):699–710.
89. Soni KA, Nannapaneni R, Hagens S. Reduction of *Listeria monocytogenes* on the surface of fresh channel Catfish Fillets by *bacteriophage Listex* P100. *Foodborne Pathogens Dis*. 2010;7(4): 427–434.
90. Hanlon GW, Denyer SP, Olliff CJ, Ibrahim LJ. Reduction in exopolysaccharide viscosity as an aid to bacteriophage penetration through *Pseudomonas aeruginosa* biofilms. *Appl Environ Microbiol*. 2001;67(6):2746–2753.
91. Knezevic P, Petrovic O. A colorimetric microtiter plate method for assessment of phage effect on *Pseudomonas aeruginosa* biofilm. *J Microbiol Methods*. 2008;74(2–3):114–118.
92. Fernández L, González S, Campelo A, M. B, reports S, u.. *Low-level Predation by Lytic Phage PhiIPLA-RODI Promotes Biofilm Formation and Triggers the Stringent Response in Staphylococcus aureus*. 2017 (nature.com).
93. Walshe GE, Pang L, Flury M, Close ME, Flintoft M. Effects of pH, ionic strength, dissolved organic matter, and flow rate on the co-transport of MS2 bacteriophages with kaolinite in gravel aquifer media. *Water Res*. 2010;44(4):1255–1269.
94. Hu J, Miyanaga K, Tanji Y. Diffusion of bacteriophages through artificial biofilm models. *Biotechnol Prog*. 2012;28(2):319–326.
95. Anany H, Chen W, Pelton R, Griffiths MW. Biocontrol of Listeria monocytogenes and *Escherichia coli* O157:H7 in meat by using phages immobilized on modified cellulose membranes. *Appl Environ Microbiol*. 2011;77(18):6379–6387.
96. Li L-L, Yu P, Wang X, et al. Enhanced biofilm penetration for microbial control by polyvalent phages conjugated with magnetic colloidal nanoparticle clusters (CNCs). *Environ Sci Nano*. 2017;4(9): 1817–1826.
97. Becker SC, Foster-Frey J, Donovan DM. The phage K lytic enzyme LysK and lysostaphin act synergistically to kill MRSA. *Fems Microbiol Lett*. 2008; 287(2):185–191.
98. Ryan EM, Alkawareek MY, Donnelly RF, Gilmore BF. Synergistic phage-antibiotic combinations for the control of *Escherichia coli* biofilms *in vitro*. *FEMS Immunol Med Microbiol*. 2012; 65(2):395–398.
99. Chaudhry WN, Concepción-Acevedo J, Park T, Andleeb S, Bull JJ, Levin BR. Synergy and order effects of antibiotics and phages in killing *Pseudomonas aeruginosa* biofilms. *PLoS One*. 2017;12(1): e0168615.

100. Megonigal J, Hines M, Visscher P. *Anaerobic metabolism: linkages to trace gases and aerobic processes*. In: *Treatise on Geochemistry*. Elsevier; 2003;8: 317–424.
101. DiChristina TJ. Effects of nitrate and nitrite on dissimilatory iron reduction by Shewanella putrefaciens 200. *J Bacteriol*. 1992;174(6): 1891–1896.
102. Yang C, Rooke JA, Cabeza I, Wallace RJ. Nitrate and inhibition of ruminal methanogenesis: microbial ecology, obstacles, and opportunities for lowering methane emissions from ruminant livestock. *Front Microbiol*. 2016;7:132.
103. Gevertz D, Telang AJ, Voordouw G, Jenneman GE. Isolation and characterization of strains CVO and FWKO B, two novel nitrate-reducing, sulfide-oxidizing bacteria isolated from oil field brine. *Appl Environ Microbiol*. 2000;66(6):2491–2501.
104. Hubert C, Voordouw G. Oil field souring control by nitrate-reducing *Sulfurospirillum* spp. that outcompete sulfate-reducing bacteria for organic electron donors. *Appl Environ Microbiol*. 2007; 73(8):2644–2652.
105. Jenneman GE, Montgomery AD, McInerney MJ. Method for detection of microorganisms that produce gaseous nitrogen oxides. *Appl Environ Microbiol*. 1986;51(4):776–780.
106. Grigoryan AA, Cornish SL, Buziak B, et al. Competitive oxidation of volatile fatty acids by sulfate- and nitrate-reducing bacteria from an oil field in Argentina. *Appl Environ Microbiol*. 2008; 74(14):4324–4335.
107. Thauer RK, Jungermann K, Decker K. Energy conservation in chemotrophic anaerobic bacteria. *Bacteriol Rev*. 1977;41(1):100–180.
108. Nemati M, Jenneman GE, Voordouw G. Mechanistic study of microbial control of hydrogen sulfide production in oil reservoirs. *Biotechnol Bioeng*. 2001;74(5):424–434.
109. Kamarisima, Hidaka K, Miyanaga K, Tanji Y. The presence of nitrate- and sulfate-reducing bacteria contributes to ineffectiveness souring control by nitrate injection. *Int Biodeterior Biodegrad*. 2018; 129:81–88.
110. Yin B, Wunch K. Combined effects of microbes and nitrate on SRB growth, souring and corrosion. In: *Corrosion 2017, New Orleans, Louisiana*. 2017.
111. Schwermer CU, Lavik G, Abed RMM, et al. Impact of nitrate on the structure and function of bacterial biofilm communities in pipelines used for injection of seawater into oil fields. *Appl Environ Microbiol*. 2008;74(9):2841–2851.
112. Videla HA, Herrera LK. Understanding microbial inhibition of corrosion. A comprehensive overview. *Int Biodeterior Biodegrad*. 2009;63(7): 896–900.
113. Greene EA, Hubert C, Nemati M, Jenneman GE, Voordouw G. Nitrite reductase activity of sulphate-reducing bacteria prevents their inhibition by nitrate-reducing, sulphide-oxidizing bacteria. *Environ Microbiol*. 2003;5(7):607–617.
114. Chen Y, Tang Q, Senko JM, et al. Long-term survival of *Desulfovibrio vulgaris* on carbon steel and associated pitting corrosion. *Corros Sci*. 2015;90: 89–100.
115. Xu D, Li Y, Song F, Gu T. Laboratory investigation of microbiologically influenced corrosion of C1018 carbon steel by nitrate reducing bacterium *Bacillus licheniformis*. *Corros Sci*. 2013;77:385–390.
116. Jia R, Yang D, Xu D, Gu T. Electron transfer mediators accelerated the microbiologically influence corrosion against carbon steel by nitrate reducing *Pseudomonas aeruginosa* biofilm. *Bioelectrochemistry*. 2017;118:38–46.
117. Gadekar S, Nemati M, Hill G. Batch and continuous biooxidation of sulphide by *Thiomicrospira* sp. CVO: reaction kinetics and stoichiometry. *Water Res*. 2006;40(12):2436–2446.
118. Yoon S, Cruz-García C, Sanford R, Ritalahti KM, Löffler FE. Denitrification versus respiratory ammonification: environmental controls of two competing dissimilatory NO3(-)/NO2(-) reduction pathways in *Shewanella loihica* strain PV-4. *ISME J*. 2015;9(5):1093–1104.
119. Erşan YÇ, Verbruggen H, De Graeve I, Verstraete W, De Belie N, Boon N. Nitrate reducing $CaCO_3$ precipitating bacteria survive in mortar and inhibit steel corrosion. *Cem Concr Res*. 2016;83:19–30.
120. Lambo AJ, Noke K, Larter SR, Voordouw G. Competitive, microbially-mediated reduction of nitrate with sulfide and aromatic oil components in a low-temperature, Western Canadian oil reservoir. *Environ Sci Technol*. 2008;42(23):8941–8946.
121. Kaster KM, Grigoriyan A, Jennneman G, Voordouw G. Effect of nitrate and nitrite on sulfide production by two thermophilic, sulfate-reducing enrichments from an oil field in the North Sea. *Appl Microbiol Biotechnol*. 2007;75(1): 195–203.
122. Voordouw G. Oil field biotechnology: should we use nitrate or nitrite to remediate souring. In: *Canadian International Petroleum Conference, Calgary, Alberta*. 2003.

123. Callbeck CM, Dong X, Chatterjee I, et al. Microbial community succession in a bioreactor modeling a souring low-temperature oil reservoir subjected to nitrate injection. *Appl Microbiol Biotechnol*. 2011; 91(3):799–810.
124. Voordouw G, Grigoryan AA, Lambo A, et al. Sulfide remediation by pulsed injection of nitrate into a low temperature Canadian heavy oil reservoir. *Environ Sci Technol*. 2009;43(24): 9512–9518.
125. Sunde E, Bodtker G, Lillebo B-L, Thorstenson T. H2S inhibition by nitrate injection on the Gullfaks field. In: *Corrosion 2004, New Orleans, Louisiana*. 2004.
126. Bødtker G, Thorstenson T, Lillebø B-LP, et al. The effect of long-term nitrate treatment on SRB activity, corrosion rate and bacterial community composition in offshore water injection systems. *J Ind Microbiol Biotechnol*. 2008;35(12): 1625–1636.
127. Alasvand Zarasvand K, Rai VR. Microorganisms: induction and inhibition of corrosion in metals. *Int Biodeterior Biodegrad*. 2014;87:66–74.
128. Schweitzer PA. *Fundamentals of Corrosion: Mechanisms, Causes, and Preventative Methods*. CRC Press; 2010:403.
129. Jayaraman A, Cheng ET, Earthman JC, Wood TK. Importance of biofilm formation for corrosion inhibition of SAE 1018 steel by axenic aerobic biofilms. *J Industrial Microbiol Biotechnol*. 1997; 18(6):396–401.
130. Abdoli L, Huang J, Li H. Electrochemical corrosion behaviors of aluminum-based marine coatings in the presence of *Escherichia coli* bacterial biofilm. *Mater Chem Phys*. 2016;173:62–69.
131. Fong JNC, Yildiz FH. Biofilm matrix proteins. *Microbiol Spectr*. 2015;3(2).
132. Limoli DH, Jones CJ, Wozniak DJ. Bacterial extracellular polysaccharides in biofilm formation and function. *Microbiol Spectr*. 2015;3(3).
133. El-Sheshtawy H, Aiad I, Osman M, Abo-ELnasr A, Kobisy A. Production of biosurfactant from Bacillus licheniformis for microbial enhanced oil recovery and inhibition the growth of sulfate reducing bacteria. *Egypt J Petroleum*. 2015;24(2): 155–162.
134. Jordan RN, Nichols EP, Cunningham AB. The role of (bio)surfactant sorption in promoting the bioavailability of nutrients localized at the solid-water interface. *Water Sci Technol*. 1999;39(7): 91–98.
135. Heinemann C, Hylckama Vlieg JE, Janssen DB, Busscher HJ, Mei HC, Reid G. Purification and characterization of a surface-binding protein from *Lactobacillus fermentum* RC-14 that inhibits adhesion of *Enterococcus faecalis* 1131. *Fems Microbiol Lett*. 2000;190(1):177–180.
136. Li B, Logan BE. Bacterial adhesion to glass and metal-oxide surfaces. *Colloids Surfaces B Biointerfaces*. 2004;36(2):81–90.
137. Finkenstadt VL, Côté GL, Willett JL. Corrosion protection of low-carbon steel using exopolysaccharide coatings from *Leuconostoc mesenteroides*. *Biotechnol Lett*. 2011;33(6):1093–1100.
138. Juzeliūnas E, Ramanauskas R, Lugauskas A, Leinartas K, Samulevičienė M, Sudavičius A. Influence of wild strain Bacillus mycoides on metals: from corrosion acceleration to environmentally friendly protection. *Electrochimica Acta*. 2006; 51(27):6085–6090.
139. Wadood HZ, Rajasekar A, Ting Y-P, Sabari AN. Role of *Bacillus subtilis* and *Pseudomonas aeruginosa* on corrosion behaviour of stainless steel. *Arabian J Sci Eng*. 2015;40(7):1825–1836.
140. Moreno DA, Ibars JR, Beech IB, Gaylarde CC. Biofilm formation on mild steel coupons by *Pseudomonas* and *Desulfovibrio*. *Biofouling*. 1993;7(2): 129–139.
141. Yun M-A, Yeon K-M, Park J-S, Lee C-H, Chun J, Lim DJ. Characterization of biofilm structure and its effect on membrane permeability in MBR for dye wastewater treatment. *Water Res*. 2006; 40(1):45–52.
142. Al-Hashem A, Carew J, Al Borno A. Screening test for six dual biocide regimes against planktonic and sessile populations of bacteria. In: *Corrosion 2004, New Orleans, Louisiana*. 2004.
143. Okabe S, Itoh T, Satoh H, Watanabe Y. Analyses of spatial distributions of sulfate-reducing bacteria and their activity in aerobic wastewater biofilms. *Appl Environ Microbiol*. 1999;65(11):5107–5116.
144. Ismail KM, Gehrig T, Jayaraman A, et al. Corrosion control of mild steel by aerobic bacteria under continuous flow conditions. *Corrosion*. 2002; 58(5):417–423.
145. Jayaraman A, Cheng ET, Earthman JC, Wood TK. Axenic aerobic biofilms inhibit corrosion of SAE 1018 steel through oxygen depletion. *Appl Microbiol Biotechnol*. 1997;48(1):11–17.
146. Landoulsi J, Kirat KE, Richard C, Féron D, Pulvin S. Enzymatic approach in microbial-influenced corrosion: a review based on stainless

steels in natural waters. *Environ Sci Technol*. 2008; 42(7):2233–2242.
147. Busalmen J, Vázquez M, de Sánchez S. New evidences on the catalase mechanism of microbial corrosion. *Electrochimica Acta*. 2002;47(12): 1857–1865.
148. Karn SK, Fang G, Duan J. Bacillus sp. acting as dual role for corrosion induction and corrosion inhibition with carbon steel (CS). *Front Microbiol*. 2017;8:2038.
149. Dubiel M, Hsu CH, Chien CC, Mansfeld F, Newman DK. Microbial iron respiration can protect steel from corrosion. *Appl Environ Microbiol*. 2002;68(3):1440–1445.
150. Soleimani S, Isgor OB, Ormeci B. Resistance of biofilm-covered mortars to microbiologically influenced deterioration simulated by sulfuric acid exposure. *Cem Concr Res*. 2013;53:229–238.
151. Chongdar S, Gunasekaran G, Kumar P. Corrosion inhibition of mild steel by aerobic biofilm. *Electrochimica Acta*. 2005;50(24):4655–4665.
152. Volkland HP, Harms H, Müller B, Repphun G, Wanner O, Zehnder AJ. Bacterial phosphating of mild (unalloyed) steel. *Appl Environ Microbiol*. 2000;66(10):4389–4395.
153. Volkland H-P, Harms H, Kaufmann K, Wanner O, Zehnder AJ. Repair of damaged vivianite coatings on mild steel using bacteria. *Corros Sci*. 2001; 43(11):2135–2146.
154. Visca P, Colotti G, Serino L, Verzili D, Orsi N, Chiancone E. Metal regulation of siderophore synthesis in *Escherichia coli* and functional effects of siderophore-metal complexes. *Appl Environ Microbiol*. 1992;58(9):2886–2893.
155. Boukhalfa H, Reilly SD, Michalczyk R, Iyer S, Neu MP. Iron(III) coordination properties of a pyoverdin siderophore produced by *Pseudomonas putida* ATCC 33015. *Inorg Chem*. 2006;45(14): 5607–5616.
156. Jayaraman A, Mansfeld FB, Wood TK. Inhibiting sulfate-reducing bacteria in biofilms by expressing the antimicrobial peptides indolicidin and bactenecin. *J Industrial Microbiol Biotechnol*. 1999; 22(3):167–175.
157. Zuo R, Wood TK. Inhibiting mild steel corrosion from sulfate-reducing and iron-oxidizing bacteria using gramicidin-S-producing biofilms. *Appl Microbiol Biotechnol*. 2004;65(6):747–753.
158. Arps PJ, Earthman JC, Xu L-C, et al. Field evaluation of corrosion control using regenerative biofilms (CCURB). In: *Corrosion 2003, San Diego, California*. 2003.
159. Örnek D, Jayaraman A, Syr B, et al. Pitting corrosion inhibition of aluminum 2024 by Bacillus biofilms secreting polyaspartate or γ-polyglutamate. *Appl Microbiol Biotechnol*. 2002; 58(5):651–657.
160. Zin' IM, Karpenko OV, Kornii SA, et al. Influence of a rhamnolipid biocomplex on the corrosion of duralumin in the case of mechanical activation of its surface. *Mater Sci*. 2016;51(5):618–626.
161. Shubina V, Gaillet L, Ababou-Girard S, et al. The influence of biosurfactant adsorption on the physicochemical behaviour of carbon steel surfaces using contact angle measurements and X-ray photoelectron spectroscopy. *Appl Surf Sci*. 2015;351:1174–1183.
162. Kuyukina MS, Ivshina IB, Korshunova IO, Stukova GI, Krivoruchko AV. Diverse effects of a biosurfactant from Rhodococcus ruber IEGM 231 on the adhesion of resting and growing bacteria to polystyrene. *Amb Express*. 2016;6(1):14.
163. Arora A, Cameotra SS, Kumar R, et al. Biosurfactant as a promoter of methane hydrate formation: thermodynamic and kinetic studies. *Sci Rep*. 2016;6(1):20893.
164. Dagbert C, Meylheuc T, Bellon-Fontaine M-N. Corrosion behaviour of AISI 304 stainless steel in presence of a biosurfactant produced by *Pseudomonas fluorescens*. *Electrochimica Acta*. 2006; 51(24):5221–5227.
165. Valle J, Da Re S, Henry N, et al. Broad-spectrum biofilm inhibition by a secreted bacterial polysaccharide. *Proc Natl Acad Sci U S A*. 2006; 103(33):12558–12563.
166. Ciandrini E, Campana R, Casettari L, et al. Characterization of biosurfactants produced by *Lactobacillus* spp. and their activity against oral *Streptococci* biofilm. *Appl Microbiol Biotechnol*. 2016;100(15):6767–6777.
167. Hentzer M, Eberl L, Givskov M. Transcriptome analysis of *Pseudomonas aeruginosa* biofilm development: anaerobic respiration and iron limitation. *Biofilms*. 2005;2(1):37–61.
168. Youssef N, Simpson DR, Duncan KE, et al. In situ biosurfactant production by bacillus strains injected into a limestone petroleum reservoir. *Appl Environ Microbiol*. 2007;73(4):1239–1247.
169. Gopinath V, MubarakAli D, Priyadarshini S, Priyadharsshini NM, Thajuddin N, Velusamy P. Biosynthesis of silver nanoparticles from *Tribulus terrestris* and its antimicrobial activity: a novel biological approach. *Colloids Surfaces B Biointerfaces*. 2012;96:69–74.

170. Ramyadevi J, Jeyasubramanian K, Marikani A, Rajakumar G, Rahuman AA. Synthesis and antimicrobial activity of copper nanoparticles. *Mater Lett*. 2012;71:114–116.
171. Padmavathy N, Vijayaraghavan R. Enhanced bioactivity of ZnO nanoparticles-an antimicrobial study. *Sci Technol Adv Mater*. 2008;9(3):035004.
172. Fazio E, Santoro M, Lentini G, Franco D, Guglielmino SPP, Neri F. Iron oxide nanoparticles prepared by laser ablation: synthesis, structural properties and antimicrobial activity. *Colloids Surfaces A Physicochem Eng Aspects*. 2016;490: 98–103.
173. Kumar N, Omoregie EO, Rose J, et al. Inhibition of sulfate reducing bacteria in aquifer sediment by iron nanoparticles. *Water Res*. 2014;51:64–72.
174. Liu T, Yin B, He T, Guo N, Dong L, Yin Y. Complementary effects of nanosilver and superhydrophobic coatings on the prevention of marine bacterial adhesion. *ACS Appl Mater Inter*. 2012;4(9): 4683–4690.
175. Lee JH, Kim Y-G, Cho MH, Lee J. ZnO nanoparticles inhibit *Pseudomonas aeruginosa* biofilm formation and virulence factor production. *Microbiol Res*. 2014;169(12):888–896.
176. Yin B, Liu T, Dong LH, Zhang L, Yin YS. Reducing microbially-influenced corrosion on aluminum by using organosiloxane sol film with silver nanocrystals. *Adv Mater Res*. 2012;463–464: 1479–1483.
177. Ikuma K, Decho AW, Lau BLT. When nanoparticles meet biofilms interactions guiding the environmental fate and accumulation of nanoparticles. *Front Microbiol*. 2015;6:591.
178. Alasvand Zarasvand K, Rai VR. Inhibition of a sulfate reducing bacterium, *Desulfovibrio marinisediminis* GSR3, by biosynthesized copper oxide nanoparticles. *3 Biotech*. 2016;6(1):84.
179. Kitching M, Ramani M, Marsili E. Fungal biosynthesis of gold nanoparticles: mechanism and scale up. *Microb Biotechnol*. 2015;8(6):904–917.
180. Eugenio M, Müller N, Frasés S, et al. Yeast-derived biosynthesis of silver/silver chloride nanoparticles and their antiproliferative activity against bacteria. *RSC Adv*. 2016;6(12):9893–9904.
181. Ramanathan R, Field MR, O'Mullane AP, Smooker PM, Bhargava SK, Bansal V. Aqueous phase synthesis of copper nanoparticles: a link between heavy metal resistance and nanoparticle synthesis ability in bacterial systems. *Nanoscale*. 2013;5(6):2300–2306.
182. Saravana Kumar P, Balachandran C, Duraipandiyan V, Ramasamy D, Ignacimuthu S, Al-Dhabi NA. Extracellular biosynthesis of silver nanoparticle using *Streptomyces* sp. 09 PBT 005 and its antibacterial and cytotoxic properties. *Appl Nanosci*. 2015;5(2):169–180.
183. Zhang H, Hu X. Biosynthesis of Pd and Au as nanoparticles by a marine bacterium *Bacillus* sp. GP and their enhanced catalytic performance using metal oxides for 4-nitrophenol reduction. *Enzyme Microb Technol*. 2018;113:59–66.
184. Labrenz, Druschel G, Thomsen-Ebert T, et al. Formation of sphalerite (ZnS) deposits in natural biofilms of sulfate-reducing bacteria. *Science*. 2000;290.
185. Kalathil S, Lee J, Cho MH. Electrochemically active biofilm-mediated synthesis of silver nanoparticles in water. *Green Chem*. 2011;13(6):1482.
186. Khan MM, Ansari SA, Lee J-H, Lee J, Cho MH. Mixed culture electrochemically active biofilms and their microscopic and spectroelectrochemical studies. *ACS Sustain Chem Eng*. 2014;2(3): 423–432.
187. Chandki R, Banthia P, Banthia R. Biofilms: a microbial home. *J Indian Soc Periodontology*. 2011; 15(2):111–114.
188. Khan MM, Kalathil S, Han TH, Lee J, Cho MH. Positively charged gold nanoparticles synthesized by electrochemically active biofilm—a biogenic approach. *J Nanosci Nanotechnol*. 2013;13(9): 6079–6085.
189. He S, Zhang Y, Guo Z, Gu N. Biological synthesis of gold nanowires using extract of *Rhodopseudomonas capsulata*. *Biotechnol Prog*. 2008;24(2):476–480.
190. Cao L, Jones AK, Sikka VK, Wu J, Gao D. Anti-icing superhydrophobic coatings. *Langmuir*. 2009; 25(21):12444–12448.
191. Choi C-H, Ulmanella U, Kim J, Ho C-M, Kim C-J. Effective slip and friction reduction in nanograted superhydrophobic microchannels. *Phys Fluids*. 2006;18(8):087105.
192. Song F, Wu C, Chen H, et al. Water-repellent and corrosion-resistance properties of superhydrophobic and lubricant-infused super slippery surfaces. *RSC Adv*. 2017;7(70):44239–44246.
193. Eduok U, Suleiman R, Gittens J, et al. Anticorrosion/antifouling properties of bacterial spore-loaded sol–gel type coating for mild steel in saline marine condition: a case of thermophilic strain of *Bacillus licheniformis*. *RSC Adv*. 2015; 5(114):93818–93830.

194. Kobayashi K, Iwano M. BslA(YuaB) forms a hydrophobic layer on the surface of *Bacillus subtilis* biofilms. *Mol Microbiol*. 2012;85(1):51–66.
195. Epstein AK, Pokroy B, Seminara A, Aizenberg J. Bacterial biofilm shows persistent resistance to liquid wetting and gas penetration. *Proc Natl Acad Sci USA*. 2011;108(3):995–1000.
196. Arnaouteli S, MacPhee CE, Stanley-Wall NR. Just in case it rains: building a hydrophobic biofilm the *Bacillus subtilis* way. *Curr Opin Microbiol*. 2016;34:7–12.
197. Guo Z, Liu T, Cheng YF, Guo N, Yin Y. Adhesion of *Bacillus subtilis* and *Pseudoalteromonas lipolytica* to steel in a seawater environment and their effects on corrosion. *Colloids Surfaces B Biointerfaces*. 2017;157:157–165.
198. Hobley L, Ostrowski A, Rao FV, et al. BslA is a self-assembling bacterial hydrophobin that coats the *Bacillus subtilis* biofilm. *Proc Natl Acad Sci USA*. 2013;110(33):13600–13605.
199. Werb M, García CF, Bach NC, et al. Surface topology affects wetting behavior of *Bacillus subtilis* biofilms. *Npj Biofilms Microbiomes*. 2017;3(1):11.
200. Arnaouteli S, Ferreira AS, Schor M, et al. Bifunctionality of a biofilm matrix protein controlled by redox state. *Proc Natl Acad Sci USA*. 2017; 114(30):E6184–E6191.
201. Wessels JG. Hydrophobins: proteins that change the nature of the fungal surface. *Adv Microb Physiology*. 1996;38:1–45.
202. Misra R, Li J, Cannon GC, Morgan SE. Nanoscale reduction in surface friction of polymer surfaces modified with Sc3 hydrophobin from *Schizophyllum* commune. *Biomacromolecules*. 2006;7(5): 1463–1470.
203. Artini M, Cicatiello P, Ricciardelli A, et al. Hydrophobin coating prevents *Staphylococcus epidermidis* biofilm formation on different surfaces. *Biofouling*. 2017;33(7):601–611.
204. Rieder A, Ladnorg T, Wöll C, Obst U, Fischer R, Schwartz T. The impact of recombinant fusion-hydrophobin coated surfaces on *E. coli* and natural mixed culture biofilm formation. *Biofouling*. 2011;27(10):1073–1085.
205. Alireza Bagherzadeh S, Alavi S, Ripmeester JA, Englezos P. Why ice-binding type I antifreeze protein acts as a gas hydrate crystal inhibitor. *Phys Chem Chem Phys*. 2015;17(15):9984–9990.
206. Obanijesu EO, Pareek V, Tade MO. Hydrate formation and its influence on natural gas pipeline internal corrosion rate. In: *SPE Oil and Gas India Conference and Exhibition*. 2010.
207. Lanoil BD, Sassen R, La Duc MT, Sweet ST, Nealson KH. Bacteria and archaea physically associated with Gulf of Mexico gas hydrates. *Appl Environ Microbiol*. 2001;67(11):5143–5153.
208. Knittel K, Boetius A, Lemke A, et al. Activity, distribution, and diversity of sulfate reducers and other bacteria in sediments above gas hydrate (Cascadia Margin, Oregon). *Geomicrobiol J*. 2003;20(4): 269–294.
209. Zhang CL, Li Y, Wall JD, et al. Lipid and carbon isotopic evidence of methane-oxidizing and sulfate-reducing bacteria in association with gas hydrates from the Gulf of Mexico. *Geology*. 2002; 30(3):239.
210. Wu Z. *Bacterial Low Temperature Survival, Ice Nucleation Proteins and Ice-associating Polymers*. 2010.
211. Xu H, Griffith M, Patten CL, Glick BR. Isolation and characterization of an antifreeze protein with ice nucleation activity from the plant growth promoting rhizobacterium *Pseudomonas putida* GR12-2. *Can J Microbiol*. 1998;44(1): 64–73.
212. Yamashita Y, Nakamura N, Omiya K, Nishikawa J, Kawahara H, Obata H. Identification of an antifreeze Lipoprotein from *Moraxella* sp. of Antarctic origin. *Biosci Biotechnol Biochem*. 2002;66(2): 239–247.
213. Kawahara H, Nakano Y, Omiya K, Muryoi N, Nishikawa J, Obata H. Production of two types of ice crystal-controlling proteins in Antarctic bacterium. *J Biosci Bioeng*. 2004;98(3): 220–223.
214. Gilbert JA, Davies PL, Laybourn-Parry J. A hyperactive, Ca^{2+} -dependent antifreeze protein in an Antarctic bacterium. *Fems Microbiol Lett*. 2005;245(1):67–72.
215. Raymond JA, Fritsen C, Shen K. An ice-binding protein from an Antarctic sea ice bacterium. *Fems Microbiol Ecol*. 2007;61(2):214–221.
216. Kawahara H, Iwanaka Y, Higa S, et al. *A Novel, Intracellular Antifreeze Protein in an Antarctic Bacterium,* Flavobacterium Xanthum. 2007.
217. Raymond JA, Christner BC, Schuster SC. A bacterial ice-binding protein from the vostok ice core. *Extremophiles*. 2008;12(5):713–717.
218. Do H, Kim S-J, Kim HJ, Lee JH. Structure-based characterization and antifreeze properties of a hyperactive ice-binding protein from the Antarctic bacterium *Flavobacterium frigoris* PS1. *Acta Crystallogr Sect D Biol Crystallogr*. 2014;70(4): 1061–1073.

219. Mangiagalli M, Bar-Dolev M, Tedesco P, et al. Cryo-protective effect of an ice-binding protein derived from Antarctic bacteria. *FEBS J*. 2017; 284(1):163–177.
220. Walker VK, Zeng H, Ohno H, et al. Antifreeze proteins as gas hydrate inhibitors. *Can J Chem*. 2015;93(8):839–849.
221. Huva E, Gordienko R, Ripmeester J, Zeng H, Walker V. The search for 'green inhibitors': perturbing hydrate growth with bugs. In: *Proceedings of the 6th International Conference on Gas Hydrates, Vancouver, British Columbia, Canada, Proceedings of the 6th International Conference on Gas Hydrates, Vancouver, British Columbia, Canada*. 2008.
222. Townson I, Walker VK, Ripmeester JA, Englezos P. Bacterial inhibition of methane clathrate hydrates formed in a stirred autoclave. *Energy & Fuels*. 2012; 26(12):7170–7175.
223. Hong SH, Marshall RT. Natural exopolysaccharides enhance survival of lactic acid bacteria in frozen dairy desserts. *JDS*. 2001;84(6):1367–1374.
224. Tamaru Y, Takani Y, Yoshida T, Sakamoto T. Crucial role of extracellular polysaccharides in desiccation and freezing tolerance in the terrestrial cyanobacterium Nostoc commune. *Appl Environ Microbiol*. 2005;71(11):7327–7333.
225. Wu Z, Kan FWK, She YM, Walker VK. Biofilm, ice recrystallization inhibition and freeze-thaw protection in an epiphyte community. *Appl Biochem Microbiol*. 2012;48(4):363–370.
226. Grote J, Krysciak D, Streit WR. Phenotypic heterogeneity, a phenomenon that may explain why quorum sensing does not always result in truly homogenous cell behavior. *Appl Environ Microbiol*. 2015;81(16):5280–5289.
227. Tan CH, Koh KS, Xie C, et al. The role of quorum sensing signalling in EPS production and the assembly of a sludge community into aerobic granules. *ISME J*. 2014;8(6):1186–1197.
228. Magnuson J. Potential role of RTA3 and GNP3 transport genes in the quorum sensing response of Candida Albicans. *Undergrad Honors Theses*. 2015.
229. Sadekuzzaman M, Yang S, Mizan M, Ha S. Current and recent advanced strategies for combating biofilms. *Compr Rev Food Sci Food Saf*. 2015; 14(4):491–509.
230. Katebian L, Gomez E, Skillman L, Li D, Ho G, Jiang SC. Inhibiting quorum sensing pathways to mitigate seawater desalination RO membrane biofouling. *Desalination*. 2016;393:135–143.
231. Yang C, Song G, Zhu Q, Liu S, Xia C. The influence of bacterial quorum-sensing inhibitors against the formation of the diatom-biofilm. *Chem Ecol*. 2016;32(2):169–181.
232. Teasdale ME, Liu J, Wallace J, Akhlaghi F, Rowley DC. Secondary metabolites produced by the marine bacterium Halobacillus salinus that inhibit quorum sensing-controlled phenotypes in gram-negative bacteria. *Appl Environ Microbiol*. 2009;75(3):567–572.
233. S. J, Cho KW, Lee HS, Rho JR, Kim TS, Mo SJ. New lactone-containing metabolites from a marine-derived bacterium of the genus Streptomyces. *J Nat Prod*. 2001;64(5):664–667.
234. Ibacache-Quiroga C, Ojeda J, Espinoza-Vergara G, Olivero P, Cuellar M, Dinamarca MA. The hydrocarbon-degrading marine bacterium *Cobetia* sp. strain MM1IDA2H-1 produces a biosurfactant that interferes with quorum sensing of fish pathogens by signal hijacking. *Microb Biotechnol*. 2013; 6(4):394–405.
235. Chen F, Gao Y, Chen X, Yu Z, Li X. Quorum quenching enzymes and their application in degrading signal molecules to block quorum sensing-dependent infection. *Int J Mol Sci*. 2013; 14(9):17477–17500.
236. Lin Y-H, Xu J-L, Hu J, et al. Acyl-homoserine lactone acylase from Ralstonia strain XJ12B represents a novel and potent class of quorum-quenching enzymes. *Mol Microbiol*. 2003;47(3): 849–860.
237. Kalia VC. Quorum sensing inhibitors: an overview. *Biotechnol Adv*. 2013;31(2):224–245.
238. Defoirdt T, Boon N, Bossier P. Can bacteria evolve resistance to quorum sensing disruption? *PLoS Pathog*. 2010;6(7):e1000989.
239. Kerekes E-B, Deák É, Takó M, et al. Anti-biofilm forming and anti-quorum sensing activity of selected essential oils and their main components on food-related micro-organisms. *J Appl Microbiol*. 2013;115(4):933–942.
240. Delago A, Mandabi A, Meijler MM. Natural quorum sensing inhibitors – small molecules, big messages. *Isr J Chem*. 2016;56(5):310–320.
241. Park SY, Kang H-O, Jang H-S, Lee J-K, Koo B-T, Yum D-Y. Identification of extracellular N-acylhomoserine lactone acylase from a Streptomyces sp. and its application to quorum quenching. *Appl Environ Microbiol*. 2005;71(5): 2632–2641.

242. Sio CF, Otten LG, Cool RH, et al. Quorum quenching by an N-acyl-homoserine lactone acylase from *Pseudomonas aeruginosa* PAO1. *Infect Immun*. 2006;74(3):1673–1682.
243. Kim M, Lee S, Park H-d, Choi S-i, Hong S. Biofouling control by quorum sensing inhibition and its dependence on membrane surface. *Water Sci Technol*. 2012;66(7):1424–1430.
244. Choo J, Rukayadi Y, Hwang J-K. Inhibition of bacterial quorum sensing by vanilla extract. *Lett Appl Microbiol*. 2006;0(0):pii: 060421055900002.
245. Kim JH, Choi DC, Yeon KM, Kim SR, Lee CH. Enzyme-immobilized nanofiltration membrane to mitigate biofouling based on quorum quenching. *Environ Sci Technol*. 2011;45(4): 1601–1607.
246. Yeon KM, Lee CH, Kim J. Magnetic enzyme carrier for effective biofouling control in the membrane bioreactor based on enzymatic quorum quenching. *Environ Sci Technol*. 2009;43(19): 7403–7409.
247. Oh HS, Tan CH, Low JH, et al. Quorum quenching bacteria can be used to inhibit the biofouling of reverse osmosis membranes. *Water Res*. 2017; 112:29–37.
248. Kim SR, Oh HS, Jo SJ, et al. Biofouling control with bead-entrapped quorum quenching bacteria in membrane bioreactors: physical and biological effects. *Environ Sci Technol*. 2013;47(2):836–842.
249. Lin Chua S, Liu Y, Li Y, et al. Reduced Intracellular c-di-GMP content increases expression of quorum sensing-regulated genes in *Pseudomonas aeruginosa*. *Front Cell Infect Microbiol*. 2017;7:451.
250. Hengge R. Trigger phosphodiesterases as a novel class of c-di-GMP effector proteins. *Philos Trans R Soc Lond Ser B, Biol Sci*. 2016;371(1707).
251. Srivastava D, Waters CM. A tangled web: regulatory connections between quorum sensing and cyclic Di-GMP. *J Bacteriol*. 2012;194(17): 4485–4493.
252. Barraud N, Hassett DJ, Hwang S-H, Rice SA, Kjelleberg S, Webb JS. Involvement of nitric oxide in biofilm dispersal of *Pseudomonas aeruginosa*. *J Bacteriol*. 2006;188(21):7344–7353.
253. Barraud N, Schleheck D, Klebensberger J, et al. Nitric oxide signaling in *Pseudomonas aeruginosa* biofilms mediates phosphodiesterase activity, decreased cyclic di-GMP levels, and enhanced dispersal. *J Bacteriol*. 2009;191(23):7333–7342.
254. Ren T, Roy R, Knowles R. Production and consumption of nitric oxide by three methanotrophic bacteria. *Appl Environ Microbiol*. 2000;66(9): 3891–3897.
255. Marshall KC, Stout R, Mitchell R. Mechanism of the initial events in the sorption of marine bacteria to surfaces. *J Gen Microbiol*. 1971;68(3): 337–348.
256. Neu TR. Microbial "footprints" and the general ability of microorganisms to label interfaces. *Can J Microbiol*. 1992;38(10):1005–1008.
257. Neu TR, Marshall KC. Microbial "footprints"—a new approach to adhesive polymers. *Biofouling*. 1991;3(2):101–112.
258. Paul JH, Jeffrey WH. Evidence for separate adhesion mechanisms for hydrophilic and hydrophobic surfaces in *Vibrio proteolytica*. *Appl Environ Microbiol*. 1985;50(2):431–437.
259. El-Kirat-Chatel S, Boyd CD, O'Toole GA, Dufrêne YF. Single-molecule analysis of *Pseudomonas fluorescens* footprints. *ACS Nano*. 2014; 8(2):1690–1698.
260. Sand W, Gehrke T. Extracellular polymeric substances mediate bioleaching/biocorrosion via interfacial processes involving iron(III) ions and acidophilic bacteria. *Res Microbiol*. 2006;157(1): 49–56.
261. Gómez-Suárez C, Pasma J, van der Borden AJ, et al. Influence of extracellular polymeric substances on deposition and redeposition of *Pseudomonas aeruginosa* to surfaces. *Microbiology*. 2002;148(4):1161–1169.
262. Stadler R, Wei L, Fürbeth W, Grooters M, Kuklinski A. Influence of bacterial exopolymers on cell adhesion of *Desulfovibrio vulgaris* on high alloyed steel: corrosion inhibition by extracellular polymeric substances (EPS). *Mater Corros*. 2010; 61(12):1008–1016.
263. Hwang G, Kang S, El-Din MG, Liu Y. Impact of an extracellular polymeric substance (EPS) precoating on the initial adhesion of *Burkholderia cepacia* and *Pseudomonas aeruginosa*. *Biofouling*. 2012; 28(6):525–538.
264. Muruve NG, Cheng YF, Feng Y, et al. Peptide-based biocoatings for corrosion protection of stainless steel biomaterial in a chloride solution. *Mater Sci Eng C*. 2016;68:695–700.
265. Hluchan V, Wheeler BL, Hackerman N. Amino acids as corrosion inhibitors in hydrochloric acid

solutions. *Mater Corrosion/Werkstoffe und Korrosion*. 1988;39(11):512–517.

266. Müller B. Amino and polyamino acids as corrosion inhibitors for aluminium and zinc pigments. *Pigment Resin Technol*. 2002;31(2):84–87.
267. Lee SY, Choi JH, Xu Z. Microbial cell-surface display. *Trends Biotechnol*. 2003;21(1):45–52.
268. Samuelson P, Wernérus H, Svedberg M, Ståhl S. Staphylococcal surface display of metal-binding polyhistidyl peptides. *Appl Environ Microbiol*. 2000;66(3):1243–1248.
269. Zuo R, Örnek D, Wood TK. Aluminum- and mild steel-binding peptides from phage display. *Appl Microbiol Biotechnol*. 2005;68(4):505–509.

CHAPTER 6

Future Perspective of Biological Inhibition of Corrosion

Protection of metal against corrosion remains a major challenge in many industries. Current methods applied for preventing the corrosion of various functional systems are physical and chemical methods. Although these methods are widely used for preventing corrosion of various functional systems, there are a number of skepticism for their use in industries. For instance, most of these methods are costly. These strategies are moderately effective in corrosion prevention and even ineffective when the bacterial biofilm is present. In case they successfully inhibit the corrosion, particularly the MIC, they may cause environmental contamination. Therefore, there is a great demand for suitable protection technologies. Recently, adequate attention dedicated to resolving the problem of corrosion by microbes. As we explained in previous chapter, there are many approaches to utilize the natural abilities of microbes for more sophisticated inhibition strategies. Although those examples illustrate the impact that microbiology can have on the problem of corrosion but note that similar to physical and chemical inhibitory strategies, microbial methods also have considerable uncertainty for a number of reasons. So, major efforts are required to refine these limitations before it becomes effective and economically viable for use in the field. In this brief chapter, mention will be made of additional areas that may—if certain developments occur—result in great economic progress and benefits as well as contributing to fundamental knowledge. Development of the potential need should not be restricted to the microbiological field; in fact, some advances in material science, chemistry and physics, may more properly treat these limitations.

Most of the inhibitory techniques we mentioned in the previous chapter are in their infancy phase and their effect on corrosion has yet to be fully examined. The majority of studies focusing on biological treatment of corrosion have been performed in vitro. So far, most of the investigations have studied the influence of these inhibitory strategies on the biofilm removal or eradication of corrosive bacteria. Although biofilm matrix is a critical target in the treatment strategies, eradication of biofilm is not always limiting the rate of biological corrosion. For instance, incomplete removal of biofilm not only cannot decrease the corrosion but also may accelerate the rate of corrosion. More basic and applied research related to bacterial activity occurring on the surfaces of metals is required. A closer look at biofilm-based bioelectrochemical reaction during the process of corrosion is another step that improves fundamental understanding of corrosion. The importance of such studies can be pointed by detection of electrochemically active biofilm; biofilm that exchanges electrons with conductive surfaces. Identification of this type of biofilms added new insight into the biocorrosion mechanisms. Understanding the effect of microorganisms on surface properties and surface composition of the film is also fundamental for improving the technology of economic value. Depending on the mineralogy of film formed on the metal surface, a film can be protective or aggressive. Although techniques such as XRD are widely used for this purpose, this technique provides limited information about physical and electronic properties of the film. Therefore, techniques that can supply more accurate analysis of the film are required. Advanced techniques such

Biological Treatment of Microbial Corrosion. https://doi.org/10.1016/B978-0-12-816108-1.00006-9

as time-of-flight secondary ion mass spectrometry(TOF-SIMS), X-ray absorption near-edge structure (XANES), and near-edge X-ray absorption fine structure spectroscopy (NXAFS) have recently been used for biofilm and biocorrosion research.[1–3] These techniques afford a real picture of the surface film without any distortion.

Much work should be done on bacterial biofilm because a better understanding of biofilm formation will facilitate the target of control strategies. How free-living planktonic bacteria switch to a sessile mode of growth, how bacteria interact with each other, and how they know it is time to leave are still some of the interesting topics in biology. Recently, research studies on microbial biofilm formed by many bacteria gave us basic information on factors and signals involved in surface sensing and attachment. For instance, it has been revealed that, in *Archaea*, certain group of regulators called the Lrs14 contributes in most of cell activities such as cell motility and EPS production or it has been shown that eDNA plays a role in the integrity of EPS.[4,5] Going forward with this type of data will shed new light for universal targeting of biofilm prevention.

Application of new and high-powered technologies appears to be of considerable importance in biofilm studies, as these techniques increase our understanding of different biological aspects of biofilm. For example, exploitation of confocal Raman microscopy and secondary ion mass spectrometry imaging provides fingerprinting of QS molecules.[6] So these techniques can reveal the fundamental role of quorum sensing–based bacterial communication during the biofilm formation or corrosion attack. Miniaturized redox sensor chips are another example of modern technology.[7] These chips that map the metabolite released by biofilm provide a better understanding of the biochemical process that occurs within biofilm. Recent advances in microscopic techniques resulted in the high-resolution imaging of biofilm. Microscopic techniques such as laser scanning microscopy (LSM), structured illumination microscopy (SIM), and optical coherence tomography (OCT) have been used in biofilmology, but these techniques are yet to be evaluated for practical application. Improvement in probes also can enhance the extraction of information related to biofilm structure. Probes can also be used for studying the biofilm diffusivity. Briandet et al.[8] investigated the diffusion and reaction of bacteriophages within biofilm by the use of fluorescence correlation spectroscopy. Additional information about biofilm structure and its diffusivity can be used for constructing the conceptual model of biofilm.

Determining whether the biological strategies work properly in a natural environment is another important aspect, which should be considered in future. Majority of studies on biological treatment of corrosion have been performed under laboratory conditions optimized for research. However, in the natural environment, microorganisms are exposed to various factors, which affect their structures and behavior. Because bacteria require definite environmental conditions for their growth, a good option for increasing the efficacy of treatment is related to the proper selection of corrosion-inhibitor microorganisms. The microorganisms should be selected based on the operating conditions. This cannot be achieved without knowing the optimum condition for the growth of selected species. Therefore, knowledge of the processes controlling bacterial behavior is an essential prerequisite for the proper exploitation of inhibitory strategies. Another strategy to improve the treatment efficacy can be obtained by control of microbial behavior in the natural environment. Research studies on the effect of resources stoichiometry on microorganisms have shown great potential of nutritious compounds on bacterial behavior.[9,10] Microbial metabolism can also be managed by genetic engineering. By using genetic engineering, scientists insert the gene of interest in bacteria and convert the organism into their minifactories, which endow a simple trait such as antibiotic, enzyme, and biosurfactant production. More investigation of the genetic and molecular biology of microorganisms and invention of more advanced techniques and instrument will improve the outcome of genetic management of microbial performance. For instance, it has been shown that, in some cases, nonlethal dose of phages can activate the genes involved in stringent response, which consequently results in enhancement of biofilm formation and increase in biofilm thickness.[11] Identification of molecular mechanisms by which phages are activating the stringent response can be used for improving the function of protective biofilm during the treatment of corrosion by regenerative biofilms. Whole-genome sequencing of microorganism, a process in which the function and loci of all genes of individual bacteria can be determined at a single time, is an approach that can help scientists to control the behavior of microbes. This technique can also be used for identification of new enzymes and antimicrobial products. The whole-genome sequence of *Bdellovibrio bacteriovorus* HD100 suggested that the vast amount of hydrolytic enzymes can be produced by these bacteria.[12]

Not only the abiotic factors but also biotic factors such as neighboring bacteria are influencing the microbial activity. So more attention to microbial interaction provides valuable information for treatment of MIC. Although biofilm in natural environments is composed of mixed-population community and microbial interaction within biofilm profoundly impacts the outcome of inhibitory treatment in nature, most of the studies are performed on single-species biofilm. Therefore, further research on mixed population biofilm that provides a more realistic picture of the in situ dynamic of biological treatment is required. Following examples are emphasizing the importance of such studies. It has been shown that single-species biofilm of *Desulfovibrio caledoniensis* accelerates corrosion, whereas dual-species biofilm of *D. caledoniensis* and *Clostridium* sp. uncultured inhibits corrosion of carbon steel.[13] Batmanghelich et al.[14] also investigated the effect of single- and dual-species biofilm of *Desulfovibrio vulgaris*, a sulfate-reducing bacterium, and *Pseudomonas aeruginosa* PAO1, a denitrifying bacterium on the corrosion of cast iron in a culture media (in situ) and in 3.5% NaCl (ex situ, pH 4). In ex situ condition, SRB biofilm was more protective than PAO1. However, in situ conditions copresence of PAO1 and SRB reduced the corrosion compared with SRB alone. The significance of microbial interaction on the successful application of inhibitory techniques can be realized in the study conducted by Zhao et al.[15] These scientists demonstrated that the production of rhamnolipids biosurfactant produced by *Pseudomonas aeroginosa* can get inhibited by H_2S. However, when rhamnolipid-producing bacteria were cocultured with *Pseudomonas stutzeri* DQI, which is a denitrifying and sulfide-removing bacteria, production of biosurfactant was resumed. In another study, Otto et al.[16] demonstrated that the fungal mycelia have a positive effect on prey-predator activity. Fungal hypha increases the accessibility of the prey to *B. bacteriovorus* 109J and, hence, allows for efficient foraging and shaping of prey populations, which cannot be seen in the absence of mycelia.

Precious omic techniques have opened new ways for more advanced analysis of microbial interactions inside the biofilm. For instance, megaproteomics determines how competitive and cooperative interactions among bacteria can influence the biofilm community. Expansion of knowledge about the effect of environmental factors on bacteria may direct scientists to find many new practical applications for economic benefits such as inhibition of microbial corrosion or biofouling. For instance, by knowing that the microscopic changes in surface roughness and surface topography can influence the bacterial surface attachments, scientists designed shark micropattern, a micropattern inspired by topography on shark skin, to prevent the bacterial adhesion.[17] These types of surfaces may provide a potential new avenue for novel nonbiocidal antibiofilm surfaces. Another example is dealing with the sensing ability of bacteria. Many bacteria can sense the changes occurring in their environment and respond to that. For instance, when some surface-attached bacteria see the blue light, they decide to leave the surface and therefore they go to their planktonic state. These data present the potential application of blue light as a nontoxic antibiofilm strategy for controlling the growth of microbes on wide varieties of material surfaces. Note that not only control of bacterial behavior but also prediction of their response is useful during the treatment. Using thermodynamics, Dolfin and Hubert[18] predicted the outcome of nitrate base treatment of oil reservoir. Outcome prediction that affords a general overview of possibilities during the application of inhibitory techniques can remarkably lessen the error approach and improve the treatment outcome.

One more topic that should be considered in the future is related to the choice of strategies. To achieve the best result, a correct choice of treatment is crucial. For instance, it is more appropriate to apply phages for preventing the MIC. If MIC is already a challenge in a system, then treatment with predatory bacteria is a better choice. Further investigation to understand the synergic effect of inhibitory strategies will help in addressing the corrosion problem. Correct selection of treatment strategies for inhibiting biological corrosion is often complicated and requires continuous monitoring of MIC. Characterization of MIC onset and offset with mixed microbial consortia will facilitate the selection of appropriate inhibitory method. Detection of corrosion redox mediators produced by certain microorganisms under stressful condition, if they are present, is a possible way of microbiologically influenced corrosion (MIC) onset determination. In a study on MIC of stainless steel caused by mixed species biofilm from equatorial seawater, Jogdeo et al.[19] revealed that the concentration of organic carbon is one of the major determinants of MIC onset. Corrosion inhibition with multifunctional approaches such as treatment with biosurfactant having antiadhesive, antibiofilm, and antimicrobial activity is recommended because these methods not only can improve the treatment efficacy but also can minimize the risk of wrong selection. Recently, Shubina et al.[20] found that a biomolecule having lipopeptide nature extracted from gram-negative bacteria cells can act as a film-forming mixed-type corrosion inhibitor and can protect the reinforced carbon steel in concrete. Because of the effective inhibitory potential of multifunctional approaches, more attention should be devoted to their development. Recent investigations are done on the application of biosurfactant for production and stabilization of nanomaterials, and also biosynthesis of gold nanoparticle with antibiofilm activity by bacteriophages inspires us for the development of more effective inhibitory techniques.[21,22]

A deeper understanding of mechanisms involved in each inhibitory technique could also be an important area of further research. Detection of new inhibitory aspects of single strategy can improve the treatment efficacy. For instance, it has been demonstrated that low doses of nitric oxide can act as adjuvant.[23] Nitric oxide reduces antibiotic tolerance of bacteria living within biofilm and makes them more susceptible to antimicrobial compounds. It should be noted that the nitric oxide is one of the intermediates produced during denitrification. Although no studies have investigated the effect of NRB on resistant properties of biofilm bacteria, such investigations can shed light on effective elimination of biofilm during the simultaneous application of NRB and biocides.

REFERENCES

1. Hua X, Marshall MJ, Xiong Y, et al. Two-dimensional and three-dimensional dynamic imaging of live biofilms in a microchannel by time-of-flight secondary ion mass spectrometry. *Biomicrofluidics*. 2015;9(3):031101.
2. Lin H, Ye C, Lv L, et al. Characterization of extracellular polymeric substances in the biofilms of typical bacteria by the sulfur K-edge XANES spectroscopy. *J Environ Sci*. 2014;26(8):1763–1768.
3. Toner B, Fakra S, Villalobos M, Warwick T, Sposito G. Spatially resolved characterization of biogenic manganese oxide production within a bacterial biofilm. *Appl Environ Microbiol*. 2005; 71(3):1300–1310.
4. Orell A, Peeters E, Vassen V, et al. Lrs14 transcriptional regulators influence biofilm formation and cell motility of Crenarchaea. *ISME J*. 2013;7(10): 1886–1898.
5. Das T, Sehar S, Manefield M. The roles of extracellular DNA in the structural integrity of extracellular polymeric substance and bacterial biofilm development. *Environ Microbiol Rep*. 2013;5(6): 778–786.
6. Baig NF, Dunham SJB, Morales-Soto N, Shrout JD, Sweedler JV, Bohn PW. Multimodal chemical imaging of molecular messengers in emerging *Pseudomonas aeruginosa* bacterial communities. *Analyst*. 2015;140(19):6544–6552.
7. Bellin DL, Sakhtah H, Zhang Y, Price-Whelan A, Dietrich LEP, Shepard KL. Electrochemical camera chip for simultaneous imaging of multiple metabolites in biofilms. *Nat Commun*. 2016;7:10535.

8. Briandet R, Lacroix-Gueu P, Renault M, et al. Fluorescence correlation spectroscopy to study diffusion and reaction of bacteriophages inside biofilms. *Appl Environ Microbiol.* 2008;74(7):2135–2143.
9. Mitra A, Flynn KJ. Predator–prey interactions: is 'ecological stoichiometry' sufficient when good food goes bad? *J Plankton Res.* 2005;27(5):393–399.
10. Yu M, Lu H, Wu D, et al. Effects of carbon-to-sulfur (C/S) ratio and nitrate (N) dosage on denitrifying sulfur cycle-associated enhanced biological phosphorus removal (DS-EBPR). *Sci Rep.* 2016;6(1):23221.
11. Fernández L, González S, Campelo AB, Martínez B, Rodríguez A, García P. Low-level predation by lytic phage philPLA-RODI promotes biofilm formation and triggers the stringent response in *Staphylococcus aureus*. *Sci Rep.* 2017;7:40965.
12. Rendulic S, Jagtap P, Rosinus A, et al. A predator unmasked: life cycle of Bdellovibrio bacteriovorus from a genomic perspective. *Science.* 2004; 303(5658):689–692.
13. Duan J, Wu S, Zhang X, Huang G, Du M, Hou B. Corrosion of carbon steel influenced by anaerobic biofilm in natural seawater. *Electrochim Acta.* 2008;54(1):22–28.
14. Batmanghelich F, Li L, Seo Y. Influence of multispecies biofilms of *Pseudomonas aeruginosa* and *Desulfovibrio vulgaris* on the corrosion of cast iron. *Corrosion Sci.* 2017;121:94–104.
15. Zhao F, Zhou J-D, Ma F, et al. Simultaneous inhibition of sulfate-reducing bacteria, removal of H2S and production of rhamnolipid by recombinant *Pseudomonas stutzeri* Rhl: applications for microbial enhanced oil recovery. *Bioresour Technol.* 2016;207: 24–30.
16. Otto S, Bruni EP, Harms H, Wick LY. Catch me if you can: dispersal and foraging of *Bdellovibrio bacteriovorus* 109J along mycelia. *ISME J.* 2017; 11(2):386–393.
17. May RM, Magin CM, Mann EE, et al. An engineered micropattern to reduce bacterial colonization, platelet adhesion and fibrin sheath formation for improved biocompatibility of central venous catheters. *Clin Transl Med.* 2015;4(1):9.
18. Dolfing J, Hubert CRJ. Using thermodynamics to predict the outcomes of nitrate-based oil reservoir souring control interventions. *Front Microbiol.* 2017;8:2575.
19. Jogdeo P, Chai R, Shuyang S, et al. Onset of microbial influenced corrosion (MIC) in stainless steel exposed to mixed species biofilms from equatorial seawater. *J Electrochem Soc.* 2017;164(9): C532–C538.
20. Shubina V, Gaillet L, Chaussadent T, Meylheuc T, Creus J. Biomolecules as a sustainable protection against corrosion of reinforced carbon steel in concrete. *J Clean Prod.* 2016;112:666–671.
21. Kiran GS, Sabu A, Selvin J. Synthesis of silver nanoparticles by glycolipid biosurfactant produced from marine *Brevibacterium casei* MSA19. *J Biotechnol.* 2010;148(4):221–225.
22. Ahiwale SS, Bankar AV, Tagunde S, Kapadnis BP. A bacteriophage mediated gold nanoparticles synthesis and their anti-biofilm activity. *Indian J Microbiol.* 2017;57(2):188–194.
23. Howlin RP, Cathie K, Hall-Stoodley L, et al. Low-dose nitric oxide as targeted anti-biofilm adjunctive therapy to treat chronic *Pseudomonas aeruginosa* infection in cystic fibrosis. *Mol Ther.* 2017;25(9): 2104–2116.

Index

A

Acid-producing bacteria
  nonorganic producers, 34
  organic acid producers, 34–35
"Aggressive" entity, 27–28
Amino acids
  carbon, 73
  fibrous proteins, 73–74
  globular proteins, 73–74
  hydrogen, 73
  nitrogen, 73
  oxygen, 73
  structure, 73, 74f
  sulfur, 73
Anode, 15
Antifreeze proteins (AFPs), 128–129
Antimicrobial producing biofilm, 121–122, 121f

B

Bacillus subtilis, 125
Bacillus surface layer protein A (BslA), 125–126
Bacteria, 30–36
  dormancy in, 91
    endospores, 76
    persisters, 91
    viable but nonculturable cell (VBNC), 91
Bacterial cell architecture
  appendages, 83–85
    flagella, 83–85
    pili, 85
  cell envelope, 85–87
    cell wall, 85–86
    cytoplasmic membrane, 85
    glycocalyx, 87, 87f
    gram-negative bacteria, 86, 86t
    gram-positive bacteria, 86, 86t
    lipopolysaccharide (LPS), 86
  cytoplasmic region, 87–88
Bacterial growth
  chemical requirement, 91–92
  physical requirement, 91–92, 92t–93t
Bacterial locomotion, 93–94
  gliding motility, 94
  sliding motility, 94
  swarming motility, 93–94
  swimming motility, 93
  twitching motility, 94
Bacterial resistance, 111–115
Bacterial signaling pathways, 91–92
  cyclic dinucleotide signaling, 93
  quorum-sensing signaling, 92–93, 93f
Biocide, 22–23
Biocides, 61t–65t
  application of, 60–65
Biofilm, 20, 27–30, 37, 39, 94, 95t
Biofilm matrix, 147–148
Bioleaching, 33–34
Biosurfactant, 122–124, 123f, 126f
Biosynthesized nanoparticles, 124–125
"Black rust" iron sulfide, 32

C

Cathode, 15
Cathodic protection methods, 18
c-di-GMP, 132, 133f
Chemical features of environment, 39
Chemical treatment, 22–23
Chemolithotrophic microorganisms
  hydrogen oxidizers, 89
  iron oxidizers, 89
  nitrifying bacteria, 89
  sulfur oxidizers, 89
Clostridia, 34–35
Coatings, application of, 20
COFS, 56–57
Colorless SOB, 34
Concrete slab, 16–17
Confocal laser scanning microscope (CLSM), 46
Copper alloys, 55, 55f
Corrosion
  basic concepts of, 15
  essentialities for, 15–16
  thermodynamics of, 11–15
Corrosion control strategies, 13–14
Corrosion effects, 3
Corrosion engineering, 23–24
  application of cathodic protection methods, 18
  application of coatings, 20
  basic concepts of corrosion, 15
  chemical treatment, 22–23
  combination of concrete slab and steel rod, 16–17
  design factor, 20–22
  essentialities for corrosion, 15–16
  materials selection, 18–20
Corrosion engineering (*Continued*)
  series and parallel corrosion process patterns, 16
  thermodynamics of corrosion, 11–15
  treating, 17–18
Corrosion impacts, 3
Corrosion inhibition
  antimicrobial producing biofilm, 121–122, 121f
  biologically produced corrosion inhibitor, 122, 123f
  biosurfactant, 122–124, 123f, 126f
  biosynthesized nanoparticles, 124–125
  corrosive agents removal, 120
  hydrophobic biofilm, 125–127, 126f–127f
  microbial antifreeze protein, 127–129, 128f, 129t
  microbial cell surface display, 134–135
  microbial footprint, 132–134, 134f
  nitrate-reducing bacteria, 115–118, 115f–116f
  predatory microorganisms, 103–115
    predatory bacteria, 103–109
  predatory phages, 109–115
    bacterial resistance, 111–115
    Desulfovibrio aespoeensis, 109
    lysogeny, 111
    narrow host range, 109–111
    phage specificity, 109
    Pseudomonas aeruginosa, 114–115
    Pseudomonas fluorescens, 109–110
    Staphylococcus lentus, 109–110
  protective layer, 121
  regenerative biofilm, 118–120
  targeting signaling molecule, 129–132
    c-di-GMP, 132, 133f
    quorum sensing inhibitors, 129–132, 130f, 131t
Corrosion inhibitors, 22–23, 148–149
Corrosion of system impacts (CSIs), 3–5, 5f
Corrosion prevention, 14
Corrosion-related bacteria (CRB), 30, 37, 45
Corrosion stop, 14–15
Corrosion system, 56
"Corrosive bacteria," 30

Note: Page numbers followed by "f" indicate figures and "t" indicate tables.

Corrosive agents removal, 120
Cost
  of corrosion and microbial corrosion, 2–9
  function of, 1–2
Counter-corrosion impacts (CCIs), 3, 5–6, 5f
CRB fuzzy modeling
  defining membership function for
    chemical features of environment, 39
    fuzzy composite functions for, 39
    physicomechanical factors, 38
CSIs. *See* Corrosion of system impacts (CSIs)
Culture-dependent methods, 42–45
Culture-independent methods, 45–46

D
Design factor, corrosion engineering, 20–22
Desulfovibrio aespoeensis, 109
Desulfovibrio caledoniensis, 149
Desulfovibrio vulgaris, 149
"Dialect alteration," 11
Diguanylate cyclase (DGC) enzymes, 132
Diversity of organisms, 26–27

E
Ecologic costs, 3–6
Ecologic impacts of corrosion, 3
Economical costs, 2–3
Eight fundamental assumptions, 37
Electrochemical corrosion, 20
Electrochemical process, 25–26
Embedded energy (EmE), 1–2

F
Fatty acids, 75–76, 75f
  long fatty acids, 75
  medium length fatty acids, 75
  membrane fluidity, 75–76
  phospholipids, 75–76
  saturated fatty acids, 75
  short fatty acids, 75
  structure, 75f
  unsaturated fatty acids, 75
Feasible corrosion management program, 26–27
Fire water rings, 4
Fit-for-service (FFS), 12–13
"Freely swimming," 27–28
Fuzzy algorithm, 37
Fuzzy composite functions
  for CRB, 39
  for PB, 41
Fuzzy logic, 36
Fuzzy probability function, 38–40

G
"Galvanic corrosion," 20
"Green" biocides, 22–23

H
Hydrogen-induced cracking (HIC), 35
Hydrophobic biofilm, 125–127, 126f–127f
Hydrotest water discharge, 6, 6f

I
Industrial biocides, 66, 66t
Ion concentration cells, 27–28
Iron, life cycle of, 11–12, 12f
Iron-oxidizing bacteria (IOB), 33–34, 34f
Iron-reducing bacteria (IRB), 34

J
Javaherdashti model for CM-MIC, 56–68

K
"Key performance indicators" (KPIs), 65–66

L
Laser scanning microscopy (LSM), 148
Life cycle of iron, 11–12, 12f
"Likelihood," 6–7
Lysogeny, 111

M
Macromolecules, 73–79
  amino acids, 73–75
  DNA, 77–78
  fatty acids, 75–76, 75f
  monosaccharides, 76, 77f
  nucleotides, 76–79, 77f
  RNA
    messenger RNA (mRNA), 78
    ribosomal RNA (rRNA), 78–79
    small RNA, 79
    transfer RNA (tRNA), 78
Magnetic bacteria, 36
  inducing corrosion, 35–36
Magnetosome formation mechanisms, 36
Magnetotactic bacteria, 35, 35f
"Mamdani minimum fuzzy implication," 37–38
Materials selection, 18–20
Mathematics, 36–38
Matrix-enclosed bacterial populations, 27
Mechanical factors of PB fuzzy modeling, 40
Microbial antifreeze protein, 127–129, 128f, 129t
Microbial cell surface display, 134–135
Microbial conditioning film, 133–134
Microbial corrosion, 6–9
Microbial footprint, 132–134, 134f
Microbial genetic material, 88
Microbial growth, 90–91
  death phase, 91
  exponential (log) phase, 90–91
  generation time, 90
  lag phase, 90
  stationary phase, 91
Microbial interaction, 96–98
  amensalism, 97–98
  commensalism, 97
  competition, 98
  mutualism, 96–97
  parasitism, 97
  predation, 97
  protocorporation, 97
Microbial metabolism, 88–89
Microbiologically influenced corrosion (MIC), 23–24, 150
  biofilm, 27–30
  culture-dependent methods, 42–45
  culture-independent methods, 45–46
  definitions, 25
  diversity of organisms, 26–27
  as electrochemical process, 25–26
  Javaherdashti model for CM-MIC, 56–68
  "myths effect" in recognition and treatment of, 53–56
  suitable environment for, 27
  "Zugzwang effect" in, 47–53
Microorganism, 89–90
  asexual reproduction, 89, 90f
  sexual reproduction, 89
Molecular microbiology methods (MMMs), 45–46, 46t
"Myths effect" in recognition and treatment of MIC, 53–56

N
Narrow host range, 109–111
Natural biocides, 22–23
Nitrate-reducing bacteria, 115–118, 115f–116f
"No Dogs Allowed" signs, 1
"Nonmetallic shell," 18–20
Nonmetals, 11
Nonorganic producers, 34

O
Optical coherence tomography (OCT), 148
Organic acid producers, 34–35
Organic molecules
  carbohydrates, 73
  lipids, 73
  nucleic acids, 73
  proteins, 73
Organisms, classification of
  algae, 81–82
  archaea, 80–81
  bacteria, 80
  classification, 80–83
  eukaryotes, 80
  fungi, 81

Organisms, classification of (*Continued*)
molds, 81
prokaryotes, 80
protozoa, 82–83
three-domain system, 80
viruses, 83
yeasts, 81

P

Parallel corrosion processes, 16, 30, 31f
Pathogenic bacteria (PB), 37
fuzzy composite functions for, 41
fuzzy modeling, 39
mechanical factors, 40
physicoelectrostatic factors, 40
Phage specificity, 109
Phenylethylamine, 129–130
Phosphodiesterase (PDE) enzymes, 132
Pipeline topography scheme, 17–18, 18f
Pit morphology, 55
Plain carbon steel, 18–20
Polyvinylpyrrolidone (PVP), 128–129
Predatory bacteria
Bacillus subtilis NCIB3610, 107f
BALOs, 103–104
biofilm, 108–109
broad-spectrum predators, 106–107
corrosion inhibitory agent, 106–109
direct invasion of cytoplasm, 103
epibiotic predation, 103
Gram-positive bacteria, 108, 108t
multiplicity of infection (MOI), 107–108
periplasmic predation, 103
wolfpack/group predation, 103–106
Predatory phages, 109–115
bacterial resistance, 111–115
Desulfovibrio aespoeensis, 109
lysogeny, 111
narrow host range, 109–111
phage specificity, 109
Pseudomonas aeruginosa, 114–115
Pseudomonas fluorescens, 109–110
Staphylococcus lentus, 109–110
Protein structure, 74
primary structure, 74
quaternary structure, 74
secondary structure, 74
tertiary structure, 74
Pseudomonas aeruginosa, 114–115, 132–134
Pseudomonas fluorescens, 109–110

Q

Quorum sensing inhibitors, 129–132, 130f, 131t

R

Reinforced concrete corrosion, 30
Reinforced concrete metallic phase, 30
Risk
of corrosion and microbial corrosion, 2–9
function of, 1–2

S

"Scientific gut feeling," 36
Series corrosion processes, 16, 16f, 30, 31f
Sessile bacteria, 27–28
"Smart" solutions, 52–53, 53f
Stainless steel, 18–20
Staphylococcus lentus, 109–110
Steel rod, 16–17
Stress corrosion cracking (SCC), 3
Structured illumination microscopy (SIM), 148
Suitable environment for MIC, 27
"Sulfate reducer," 54
Sulfate-reducing bacteria (SRB), 17, 32–33, 33t
Sulfur-oxidizing bacteria (SOB), 34, 54

T

Temenos, 30
Thermodynamics of corrosion, 11–15
Thiobacillus errooxidans, 34
Thiobacillus thiooxidans, 56

X

X-ray absorption near-edge structure (XANES), 147–148

Z

Zugzwang effect, 12–13, 14f, 23, 47–53, 52f

FSC
www.fsc.org
MIX
Papier | Fördert
gute Waldnutzung
FSC® C083411

Kolibri 360 GmbH
Ferdinand-Jühlke-Straße 7
99095 Erfurt, Deutschland
produktsicherheit@kolibri360.de